CIVIL ENGINEERING CONTRACTUAL PROCEDURES

CIVIL ENGINEERING
CONTRACTUAL PROCEDURES

Allan Ashworth

 LONGMAN

Addison Wesley Longman Limited
Edinburgh Gate
Harlow
Essex CM20 2JE
United Kingdom
and Associated Companies throughout the world

*Published in the United States of America
by Addison Wesley Longman, New York*

First published 1998

ISBN 0 582 25127 3

British Library Cataloguing-in-Publication Data
A catalogue record for this book is available from the British Library

Library of Congress Cataloging-in-Publication Data
A catalog entry for this book is available from the Library of Congress

Set by 35 in 10/12pt New Baskerville
Produced through Longman Malaysia, PP

This book is dedicated to a friend and colleague who
died after a short illness in 1997, EurIng Roger C.
Harvey, BSc (Eng), PhD, CEng, FICE, FIStructE, MRINA.
He was a former lecturer in Civil Engineering, Queen Mary
College, University of London; Professor and Head of Civil
Engineering and Dean of Engineering at Sunderland Polytechnic
and Her Majesty's Inspector (HMI) with the Department for Education.

CONTENTS

PREFACE

The aim of this book is to provide those who are employed on civil engineering projects with some information about the contractual procedures, legislation and practices that are employed in the civil engineering industry. Whilst there are a number of books about building contracts, there are relatively few that relate to contractual procedures in civil engineering. The book is divided into six sections covering the different aspects of procurement and contracts. These are as follows:

Part One considers the principles associated with law and the nature of the English legal system. Its particular emphasis is on contract law and the settlement of disputes when things go wrong on a civil engineering project.

Part Two is about procurement and the different issues associated with the selection of contractual arrangements for works of civil engineering construction.

Part Three provides background information on the way that the industry and projects are organised and the different sorts of personnel and professions involved.

Part Four examines the clauses that are covered in the Institution of Civil Engineers Conditions of Contract (sixth edition).

Part Five considers briefly some other Conditions of Contract that are used on civil engineering projects.

The Appendices provide a brief discussion on some of the main topics associated with the contractual administration of civil engineering projects.

The book has been prepared for students on undergraduate and higher technician courses in civil engineering. The understanding of contractual procedures is essential for those who choose to practise civil engineering. Often they are not given the importance they deserve in the education process. Hopefully, this book will seek to address this issue. It will also be of interest and assistance to students on courses of other related disciplines who need to gain an understanding of the practices and procedures used for the administration of civil engineering contracts. Those employed in practice will also find it a useful reference book.

This book originated from its sister publication, *Contractual Procedures in the Construction Industry* (3rd edn, Addison Wesley Longman, 1996). This has received good recommendations from many individuals and is now on reading lists of a large number of undergraduate and higher technician courses. It has been suggested by some that the book is perhaps biased in favour of building. This is a fair comment, and I have now sought to redress this bias by preparing a text that relates to civil engineering contractual procedures.

Allan Ashworth
York 1997

PART ONE

CONTRACT LAW

CHAPTER 1

THE ENGLISH LEGAL SYSTEM

Introduction

There are many individuals within the construction industry who will, at some time in their careers, become professionally involved in either litigation or arbitration. The laws which are applied in the construction industry are both:

- general, e.g. English legal system
- specialist, e.g. ICE Conditions of Contract

They are general in the sense that they embrace the tenets of law appropriate to all legal decisions, and are specialist because the interpretation of construction contracts and documents requires a particular knowledge and understanding of the construction industry. It should be noted, however, that the interpretation and application of law will never be contrary to or in opposition to the established legal principles and precedents found elsewhere. It is appropriate at this stage to consider briefly the framework of the English legal system.

The nature of law

Law, in its legal sense, may be distinguished from scientific law or the law of nature and from the rules of morality.

- Scientific laws are not man-made and are not therefore subject to change.
- Morality – It is less easy to draw a distinction between legal rules and moral precepts. It may be argued, for example, that the legal rules follow naturally from a correct moral concept. The difference between the two is, perhaps, that obedience to law is enforced by the state whereas morals are largely a matter of conscience and conduct.

The laws of a country are, however, to some extent an expression of its current morality, since laws can generally only be enforced by a common consensus. Law, therefore, may be appropriately defined as a body of rules for the guidance of human conduct but which may be enforced by the authorities concerned.

Table 1.1 Classification of law

Law	Procedure	Result
Criminal law	Prosecution for a criminal offence	Punish the defendant
Civil law	Sue for a civil wrong	Obtain some form of compensation or other benefit

Classification of law

Law is an enormous subject and some specialisation is therefore essential. A complete classification system would require a very detailed chart. Essentially, the basic division in the English legal system is the distinction between criminal and civil law. Usually, the distinction will be obvious. Alternative methods of classification are to subdivide the offences that are committed against persons, property or the state under these headings. Laws may also be classified as either public or private. Public law is primarily concerned with the state itself. Private law is that part of the English legal system which is concerned with the rights and obligations of individuals.

Sources of English law

Every legal system has its roots, the original sources from which authority is drawn. The sources of English law can be categorised in the following ways.

Custom

In the development of the English legal system the common law was derived from the different laws associated with the different parts of the country. These were adapted to form a national law common to the whole country. Since the difference between the regions stemmed from their different customary laws, it is no exaggeration to say that custom was the principal original source of the common law. The term 'custom' has three generally accepted meanings:

- general custom – accepted by the country at large
- mercantile custom – principles established on an international basis
- local custom – applicable only to certain areas within a country.

The following conditions must be complied with before a local custom will be recognised as law:

- The custom must have existed from 'time immemorial'. The date for this has been fixed as 1189.
- The custom must be limited to a particular locality.

- The custom must have existed continuously.
- The custom must be a reasonable condition in the eyes of the law.
- The custom must have been exercised openly.
- The custom must be consistent with, and not in conflict with, existing laws.

In some countries the writings of legal authors can form an important source of law. In England, however, because of tradition, such writings have in the past been treated with comparatively little respect and so are rarely cited in the courts. This general rule has always been subject to certain exceptions and there are therefore 'books of authority' which are treated as almost equal to 'precedents'. Many of these books are very old, and in some cases date back to the twelfth century.

Legislation

The majority of new laws are made in a documentary form by way of an Act of Parliament. Statute has always been a source of English law and by the nineteenth century it rivalled decided cases as a source. If statute and common law clash, the former will always prevail, since the courts cannot question the validity of any Act. The acceptance by the courts of Parliament's supremacy is entirely a matter for history. Today it is the most important new source of law because:

- The complex nature of commercial and industrial life has necessitated legislation to create the appropriate organisations and legal framework.
- Modern developments such as drugs and the motor car have necessitated legislation to prevent their abuse.
- There are frequent changes in the attitudes of modern society, such as that relating to women, and the law must thus keep in step with society.

Before a legislative measure can become law it must undergo an extensive process.

- The measure is first drafted by civil servants who present it to the House of Commons or the House of Lords as a Bill.
- The various clauses of the Bill will already have been accepted and agreed by the appropriate government department prior to its presentation.
- Before the Bill can become an Act of Parliament it must undergo five stages in each house:
  1. First reading – the Bill is introduced to the House.
  2. Second reading – a general debate takes place upon the general principles of the measure.
  3. Committee stage – each clause of the Bill is examined in detail.
  4. Report stage – the House is brought up to date with the changes that have been made.
  5. Third reading – only matters of detail are allowed to be made at this stage.

The length of time necessary for the Bill to pass through these various stages depends upon the nature and length of the Bill and how politically controversial it is. Once the Bill has been approved and accepted by each House, it then needs the Royal Assent for it to become law.

A public Bill is legislation which affects the public at large and applies throughout England and Wales. It should be noted at this point that Scottish law, although similar, is different from English law. A private Bill is legislation affecting only a limited section of the population, for example, in a particular locality. A Private Member's Bill is a public Bill introduced by a back-bench Member of Parliament as distinct from a public Bill which is introduced by the government in power.

Delegated legislation arises when a subordinate body makes laws under specific powers from Parliament. These can take the form of:

• orders in council
• statutory instruments
• bye-laws.

Whilst these are essential to the smooth running of the nation, the growth of delegated legislation can be criticised, because law making is transferred from the elected representatives to the minister, who is, in effect, the civil servants. The validity of delegated legislation can be challenged in the courts as being *ultra vires*, i.e. beyond the powers of the party making it, thus making it void. The judicial safeguard depends on the parent legislation, i.e. the Act giving the powers. Often this is extremely wide and such a restraint may therefore be almost ineffectual.

All legislation requires interpretation. The object of interpretation is to ascertain Parliament's will as expressed in the Act. The courts are thus – at least in theory – concerned with what is stated and not with what they believe Parliament intended. A large proportion of cases reported to the House of Lords and the High Court involve questions of statutory interpretation and in many of these the legislature's intention is impossible to ascertain because it never considered the question before the court. The judge must then do what he thinks Parliament would have done had it considered the question.

Since Britain's entry into the European Community (now the European Union (EU)) on 1 January 1973, it has been bound by Community law. All existing and future Community law which is self-executing is immediately incorporated into English law. A self-executing law therefore takes immediate effect and does not require action by the United Kingdom legislature.

Case law

Case law is often referred to as judicial precedent. It is the result of the decisions made by judges who have laid down legal principles derived from circumstances of the particular disputes coming before them. Importance is attached to this form of law in order that some form of consistency in application can be achieved in practice. The doctrine of judicial precedent is known as *stare decisis,* which literally means 'to stand upon decisions'. In practice, therefore, a judge, when trying a case, must always look back to see how previous judges dealt with similar cases. In looking back the judge will expect to discover those principles of law which are relevant to the case now being decided. The decision made will therefore seek to be in accordance with the already established principles of law and may, in turn, develop those principles further. One fact which should be noted is that the importance of case law is governed by the status of the court which decided the case. The cases decided in a higher court will take precedence over the judgements in a lower court.

The main advantages claimed for judicial precedent are:

- certainty – because judges must follow previous decisions, a barrister can usually advise a client on the outcome of a case
- flexibility – it is claimed that case law can be extended to meet new situations, thereby allowing the law to adjust to new social conditions

A direct result of the application of case law is that these matters must be properly reported and published and should be readily available for all future users. Consequently, there is now available within the English legal system an enormous collection of law reports stretching back over many centuries. Within the construction professions a number of different firms and organisations now collate and publish law reports which are relevant to this industry. Computerised systems are also available to allow for rapid access and retrieval from such reports.

It is not the entire decision of a judge that creates a binding precedent. When a judgement is delivered, the judge will give the reason for the decision. This is known as the *ratio decidendi,* and is a vital part of case law. It is the principle which is binding on subsequent cases that have similar facts in the same branch of law. The second aspect of judgements, *obiter dicta,* are things said 'by the way', and these do not have to be followed. Although the facts of a case appear similar to a binding precedent, a judge may consider that there is some aspect or fact which is not covered by the *ratio decidendi* of the earlier case. The judge will therefore 'distinguish' the present case from the earlier one which created the precedent.

A higher court may also consider that the *ratio decidendi* set in a lower court is not the correct law to be followed. When another case is argued on similar facts, the higher court will overrule the previous precedent and set a new precedent to be followed in future cases. Such a decision does not affect the parties in the earlier case, unlike a decision that is reversed on appeal.

Finally, a superior court may consider that there is some doubt as to the standing of a previous principle, and it may disapprove but not expressly overrule the earlier precedent.

Examples

The following are some examples of how the above sources of English law are appropriate to the construction industry:

Custom   – Right to light
            – Right of way
Legislation – Highways Act 1980
            – Town and Country Planning Act 1990
            – Local Government Act 1988
            – Control of Pollution Act 1974 (See also Table of Statutes)
Cases   – Bottoms v Lord Mayor of York 1812
            – A E Farr v The Admiralty 1953
            – Howard Marine v Ogden and Sons 1978

European Union law

A completely new source of English law was created when Parliament passed the European Communities Act 1972. Section 2(1) of the Act provides that:

> All such rights, powers, liabilities, obligations and restrictions from time to time created or arising by or under the Treaties, and all such remedies and procedures from time to time provided for by or under the Treaties, as in accordance with the treaties are without further enactment to be given legal effect or used in the United Kingdom shall be recognised and available in law, and be enforced, allowed and followed accordingly . . .

The effect of this section is that all United Kingdom courts have to recognise European Union law, whether it comes directly from treaties or other Community legislation. As soon as the 1972 Act became law, some aspects of English law were changed to bring them into line with European Union law.

There are several institutions to implement the work of the European Union. These include the European Parliament, the Council of Ministers, the Commission and the European Court of Justice.

The courts

There are a number of different courts in which civil actions may be tried. Cases are first heard in either the High Court or the County Court and, should an appeal be necessary, the matter is brought to the Court of Appeal. Where the matter is still not resolved, then it is brought to the House of Lords. Technical cases may be heard in the Official Referee's court. A further alternative is to allow the dispute to be settled through arbitration or alternative dispute resolution procedures (see Chapter 3).

County Court

There are approximately 340 districts in England and Wales in which a County Court is held at least once a month. These are divided into 60 circuits, each with its own judge. A County Court can hear almost all types of civil cases. Its jurisdiction is limited to actions in contract and tort up to sums of £5,000; actions for the recovery of land where the net annual value for rating does not exceed £1,000; equity proceedings where the sum involved does not exceed £30,000 and some bankruptcy claims. These sums are kept under review. The main advantages claimed for County Courts are their lower costs and shorter delays before the case comes to trial. Two similar advantages are claimed for arbitration. The county court comprises a circuit judge assisted by a registrar who is a solicitor of seven years' standing. The latter mainly performs administrative tasks but, with the leave of the judge, may hear actions in which the defendant does not appear, admits liability or where the claim does not exceed £2,000.

In order to reduce litigation coming before the County Court, and to enable those with very small claims to bring an action without fear of excessive court costs, a 'small claims' procedure is now available.

The High Court

The High Court hears all the more important civil cases. It is the lower half of the Supreme Court of Judicature and was brought into being under the Judicature Acts 1873–1875. It comprises three divisions which all have equal competence to try any actions, according to the pressure of work, although certain specific matters are reserved for each of them.

The Queen's Bench Division (known as QBD) deals with all types of common law work, such as contract and tort. This is the busiest division. Matters concerning the construction industry usually come to this High Court. The division is headed by the Lord Chief Justice and there are about forty to fifty lesser judges. These are known as puisne (pronounced 'puny') judges. There are two specialist courts within QBD. The Commercial Court hears major commercial disputes, usually in private, with the judge hearing the case in the more informal role of an arbitrator. The Admiralty Court hears maritime disputes.

The Chancery Division deals with such matters as trusts, mortgages, deeds, and land, taxation and partnership disputes. The division, whilst nominally headed by the Lord Chancellor and the Master of the Rolls, is actually run by a vice-chancellor, with the help of about ten to twelve lesser judges.

The Family Division deals with matters of family disputes such as probate and divorce. This division is headed by a President and three lesser judges.

The High Court normally sits at The Strand in London but there are fifteen other towns to which judges of the High Court travel to hear common law claims.

The Court of Appeal

Once a case has been heard, either party may consider an appeal. This means that the case is transferred to the Court of Appeal, where three judges usually sit to form a court. The High Court has the right to refuse an appeal. In civil appeals the appellant has six weeks from the date of judgement in which to give the Court of Appeal formal notice of appeal. The appellant must specify the exact grounds on which the appeal is based and on which the lower court reached an 'incorrect' decision.

The Civil Division hears appeals on questions of law and of fact, rehearsing the whole of the evidence presented to the court below and relying on the notes made at the trial. If the appeal is allowed, the court may reverse the decision of the lower court, or amend it, or order a retrial. It can hear appeals about the exercise of discretion, for example, discretion as to costs.

Most appeals are heard by three judges, although some (e.g. appeals from County Court decisions) can be heard by only two judges. Decisions need not be unanimous. The head of the court is the Master of the Rolls, perhaps the most influential appointment in our legal system.

The House of Lords

Appeals from decisions of the Court of Appeal are made to the House of Lords, although leave must first be obtained to do so and this is sparingly given. Permission is only given if the appeal is of general legal importance. The court used to sit in the Chamber of the House of Lords but since 1948 it has usually sat as an Appellate Committee in a committee room in the Palace of Westminster. The normal rule is that a case can only go to the House of Lords after it has been heard by the Court of Appeal, so the case progresses slowly up the judicial hierarchy. In exceptional circumstances it is possible to leap-frog over the Court of Appeal, but this is rarely done. There must be at least three judges for a committee to be quorate, although in practice appeals are heard by five judges.

All of the courts must apply statute law in reaching their decisions and, in general, the lower courts are bound by the decisions of the higher courts. In practice, the law resulting from a case to the House of Lords can only be changed by an Act of Parliament.

Arbitration

Arbitration is an alternative to litigation in the courts, and is widely used for the settlement of disputes which involve technical or commercial elements. The tribunal is chosen by the parties concerned, and the powers of the arbitrator. Arbitration is more fully explained in Chapter 3.

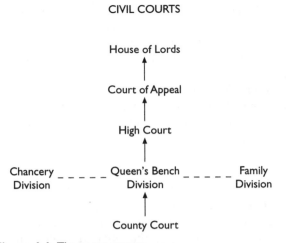

CIVIL COURTS

House of Lords

Court of Appeal

High Court

Chancery Division ---- Queen's Bench Division ---- Family Division

County Court

Figure 1.1 The court system

The lawyers

Solicitors

These are the lawyers that the public most frequently meets and, as such, are the general practitioners of the legal profession. A solicitor's work falls into the two main categories of court work and non-court work. The latter accounts for about three-quarters of their business. A solicitor operates in many ways like a businessman, with an office to run, clients to see and correspondence to be answered. Traditionally, property (conveyancing and probate) has been one of the main fee earners for solicitors.

The Solicitors Act 1974 gave solicitors three monopolies: of conveyancing, of probate, and of suing and starting court proceedings. The conveyancing monopoly was, however, significantly eroded by the Administration of Justice Act 1985. This allowed for licensed conveyancers to do this work.

The Law Society is the solicitors' professional body and as such it has two roles: firstly, to act as the representative body of solicitors, and secondly, to ensure that proper professional standards are maintained and that defaulting solicitors are disciplined. It lays down rules on professional conduct, the most important of which prohibits touting for business or unfair attraction of business. Advertising of solicitors' services is now allowed in the press and on the radio, but they must be very careful what they say about themselves. They cannot claim, for example, that they do a better job than another firm of solicitors.

Barristers

There are considerably fewer barristers than solicitors. They are specialist advocates and the specialist advisers of the legal profession. About 10 per cent of the Bar is made up of Queen's Counsel – i.e. 'those who have taken silk'. Whilst solicitors may appear in the lower courts, barristers almost have a monopoly over appearing in the higher courts. Some of their work is non-court work, such as advising on difficult points of law or on how a particular case should be conducted.

Barristers cannot form partnerships with other barristers; several barristers will, however, share a set of rooms, known as chambers, and employ clerical staff and a clerk between them. However, they do not share their earnings as do the partners in a firm of solicitors.

Future considerations

Law is sometimes seen as a question of how far you can afford to go rather than how good your case is. Lord Wolf is currently heading a commission to help remedy the huge problems of cost, delay, complexity and inequality in the civil justice system. The key is to recognise that justice is not an abstract quality. It has to be proportionate, within the means of the parties and expeditious. Many people are denied access to the courts because the costs involved are disproportionate to their claims. Lord Wolf is suggesting that in future, judges should become trial managers able to dictate the pace of legal cases. Attitudes must change but the powers to encourage settlement or to strike out unworthy cases must also be provided. Judges should also be able to encourage litigants to look at other ways of settling disputes, such as mediation or alternative dispute resolution (see Chapter 3). These procedures encourage early settlement. Judges should also to be able to give summary judgement, leaving only the core of the dispute to go to trial. The key proposals include:

- small claims court expanded to take claims up to £3,000
- fast-track cases up to £10,000, including personal injury claims with capped costs and fixed hearings
- new multi-track for cases above £10,000 providing 'hands on' management teams by judges for heaviest cases
- new post of Head of Civil Justice to run all civil courts as a single system. Post to be filled by a senior judge
- better use of technology, with laptop computers for all judges and video conference facilities
- incentives for early settlement, including 'plaintiff's offer' and referral to alternative dispute procedures
- solicitors to inform clients of charges as the bill for legal services mounts.

The current system is often excessive, disproportionate and unpredictable. The delay is frequently unreasonable. In 1994, for example, High Court cases took on average 163 weeks in London just to reach trial. Outside of London it was even longer at 189 weeks. In the County Courts the typical figure was 80 weeks or one and a half years.

The report recommends a three-track civil system with a single entry point, headed by a senior judge in a new post of Head of Civil Justice. A key point of the proposed system is to encourage settlement. Both parties will be able to make offers to settle at any stage, relating to the whole case or to just one of the issues involved.

Every year almost 300,000 writs are lodged with the Lord Chancellor's Office. Only a fraction, less than 3 per cent, ever come to trial.

Some legal jargon

abate	To reduce or make less
actus reus	The guilty act
ad valorem	According to the value. For example, stamp duty on sale of land is charged according to the price paid
affidavit	A written statement to be used as evidence in court proceedings
ancient lights	Windows which have had an uninterrupted access of light for at least twenty years. Buildings cannot be erected which interfere with this right of light.
attestation	The signature of a witness to the signing of a document by another person
burden of proof	The obligation of proving the case
caveat emptor	Let the buyer beware
certiorari	An order of the High Court to review and quash the decision of the lower court which was based on an irregular procedure
chattels	All property other than freehold real estate
Consensus ad idem	Agreement to the same idea
consideration	Where a person promises to do something for another it can only be enforced if the other person gave or promised to give something of value in return. Every contract requires consideration
costs	The expenses relating to legal services
counterclaim	When a defendant is sued any claim against the plaintiff may be included, even if it arises from a different matter
custom	An unwritten law dating back to time immemorial
defendant	A person who is sued or prosecuted, or who has any court proceedings brought against him
enactment	An Act of Parliament, or part of an Act
estoppel	A rule which prevents a person denying the truth of a statement or the existence of facts which another person has been led to believe

ex parte	An application to the court by one party to the proceedings without the other party being present
expert witness	One who is able to give an opinion on a subject. This is an exception to the rule that a witness must only tell the facts
fieri facias	A court order to the sheriff requiring the seizure of a debtor's goods to pay off a creditor's judgement
frustration	A contract is frustrated if it becomes impossible to perform because of a reason that is beyond the control of the parties. The contract is then cancelled
good faith	Honesty
goodwill	The whole advantage, wherever it may be, of the reputation and connection of a firm
in camera	When evidence is not heard in open court.
injunction	A court order requiring someone to do, or to refrain from doing, something
judicial review	An application made to the Divisional Court when a lower court or tribunal has behaved incorrectly
limitation	Court proceedings must begin within a limitation period. Different periods exist for different types of claim
liquidated sum	A specific sum, or a sum that can be worked out as a matter of arithmetic
liquidator	A person who winds up a company
moiety	One-half
obiter dictum	A statement of opinion by a judge which is not relevant to the case being tried. It is not of such authority as if it had been relevant to the case being tried
official referee	A layman appointed by the High Court to try complex matters in which he is a specialist
out of court	Settlement is reached prior to court proceedings commencing
plaintiff	Person who sues
plc	A public limited company. Most used to call themselves 'Ltd' but changed this when UK company law was brought into line with EC law in 1981
pleadings	Formal written documents in a civil action. The plaintiff submits a statement of claim and the defendant a defence
pre-trial review	Preliminary meeting of parties in a County Court action to consider administrative matters and what agreement can be reached prior to the trial

quantum meruit	As much as has been earned
ratio decidendi	The reason for a judicial decision. A statement of legal principle in a *ratio decidendi* is more authoritative than if in an *obiter dictum*
res ipsa loquitor	The matter speaks for itself
retrospective legislation	An Act that applies to a period before the Act was passed
seal	This used to be an impression in a piece of wax on a document. Now a small red sticky label is used instead. The absence of the seal will not invalidate the document, since to constitute a sealing neither wax, wafer, a piece of paper or even an impression is necessary
sine die	Indefinitely
special damage	Financial loss that can be proved
specific performance	When a party to a contract is ordered to carry out their part of the bargain. Only usually ordered where monetary damages would be an inadequate remedy
stare decisis	To stand by decided matters. Alternative name for the doctrine of precedent
statute	An Act of Parliament
statutory instrument	Subordinate legislation made by the Queen in Council or a minister, in exercise of a power granted by statute
stay of proceedings	When a court action is stopped by the court
subpoena	A court order that a person attends court, either to give evidence or to produce documents
uberrimae fidei	Of utmost good faith.
unenforceable	A contract or other right that cannot be enforced because of a technical (legal) defect
vicarious liability	When one person is responsible for the actions of another because of their relationship
void	Of no legal effect
voidable	Capable of being set aside
with costs	The winner's cost will be paid by the loser
writ	The document that commences many High Court actions

CHAPTER 2

LEGAL ASPECTS OF CONTRACTS

Introduction

A number of Acts of Parliament affect construction contracts, although it is only during this present century that they have begun to play any significant part. Historically, the law of contract has evolved by judicial decisions, so that there now exists a body of principles which apply generally to all types of construction contract. These principles have been accepted on the basis of proven cases that have been brought before the courts.

Construction contracts are usually made in writing, using one of the standard forms that are available. The use of a standard form provides many advantages, and although standard forms are not mandatory in practice their use should be encouraged in all possible circumstances. It is important to remember, however, that the making of a contract does not necessarily require any special formality. A binding contract could be made by an exchange of letters between the parties, rather than by signing an elaborate printed document. On some occasions a binding contract could be made by a gentlemen's agreement, i.e. by word of mouth. Nevertheless, there are many practical reasons why construction contracts for all but the simplest projects should be made using an approved and accepted form of contract and should be in writing. In a well drafted contract the clear intentions of the parties involved are able to be seen and understood.

Definition of a contract

A contract has been defined by Sir William Anson as 'a legally binding agreement made between two or more parties, by which rights are acquired by one or more to acts or forbearances on the part of the other or others'. The essential elements of this definition are as follows:

- **legally binding**: Not all agreements are legally binding. In particular there are social or domestic arrangements which are made without any intention of creating legal arrangements.
- **two or more parties**: In order to have an agreement there must be at least two parties. In law one cannot make bargains with oneself.

- **rights are acquired**: An essential feature of a contract is that legal rights are acquired. One person agrees to complete part of a deal and the other person agrees to do something else in return.
- **forbearances**: To forbear is to refrain from doing something. There may thus be a benefit to one party to have the other party promise not to do something.

Agreement

The whole basis of the law of contract is agreement. Specifically, a contract is an agreement bringing with it obligations which are able to be enforced in the courts if this becomes necessary. Most of the principles of modern contract date from the eighteenth and nineteenth centuries. The concept of a contract at that time was of equals coming together to bargain and reach agreement, which they would wish to be upheld by the courts. Whilst it is still true that individuals come together to form agreements, it should be recognised that many contracts are formed between parties who are not equals in any way, even where the law may pretend that they are. A major criticism of contract law in recent years has been that the wealthy, experienced and legally advised corporations have been able to make bargains with many people who are themselves of limited resources and poorly legally represented. Because of this, the law of contract has gradually moved away from a total commitment to enforce, without qualification, any agreement which has the basic elements of a contract. In particular, Parliament has introduced statutes which are often designed to protect relatively weak consumers from businessmen with greater bargaining power. Despite this, the courts are still reluctant to set aside an agreement having all of the elements of a contract and in this respect follow their nineteenth-century predecessors.

The elements of a contract

Capacity

In general, every person has full legal powers to enter into whatever contracts they might choose. There are, however, some broad exceptions to this general rule. Infants and minors, that is anyone under the age of 18 years as set out in the Family Law Reform Act 1969, cannot contract other than in certain circumstances, such as, for example, for necessaries or benefits. Persons of an unsound mind, as defined in the Mental Health Act 1959, can never make a valid contract. Other persons of an unsound mind and those unbalanced by intoxication are treated alike. Their contracts are divided into two types, those for necessary goods (the situation with minors above) and other contracts where the presumption is one of validity.

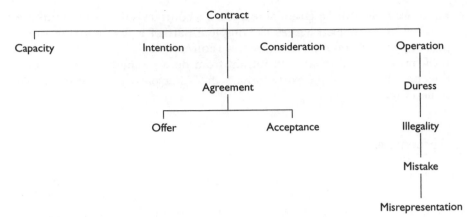

Figure 2.1 The elements of a contract

Corporations are legal entities created by a process of law. A company can only contract on matters falling within its objects clause, and since the records of companies are matters of public record available for inspection at Companies House it used to be the case that a company could not have a contract enforced against it if it lay outside its objects clause. The presumption was that those entering into contracts with a company knew or ought to have known the contents of the objects clause. Consequently anyone making an *ultra-vires* (outside-the-powers) contract with a company only had themselves to blame. On entry into the European Community (now the European Union) in 1973, this *ultra-vires* doctrine had to be revised, since it is not followed in the other countries of the Community.

Intention to create legal relations

As mentioned earlier, it cannot be assumed that an enforceable contract exists merely because there is an agreement. English law requires that the parties to a contract actually intended to enter into legal relations. These are relations actionable and enforceable in the courts. If it can be demonstrated that no such intention existed then the courts will not intervene, despite the presence of both agreement and consideration. In commercial agreements the courts presume that the parties do intend to enter into legal relations. (This is different from social, family and other domestic agreements, where the general rule is that the courts presume that there is no intention to enter into legal relations.) These are the general rules and it is also possible to demonstrate the opposite intention.

In contract law we need to know what we have agreed. It is possible for two parties to use words which are susceptible to interpretation in different ways, so that they do not have the same idea in mind when they agree. It is important that both parties have agreement to the same idea *(consensus ad idem)*. A classic case to which students should refer is *Raffles v. Wichelhaus (1864)*.

Offer and acceptance

The basis of the contract is agreement and this is composed of two parts: offer and acceptance. In addition, conditions are generally required by law (in all but the simplest of contracts) to make the offer and acceptance legally binding.

The offer

An offer must be distinguished from a mere attempt to negotiate. An offer, if it is accepted, will become a binding contract. An invitation for contractors to submit tenders is an invitation to firms to submit offers for doing the work. The invitation often states that the employer is not bound to accept the lowest or any tender or to be responsible for the costs incurred.

An offer may be revoked by the person who made it, at any time before it is accepted. Thus a contractor may submit a successful tender in terms of winning the contract. However, the contractor may choose to revoke this offer prior to formal acceptance.

Tenders for construction works do not remain on offer indefinitely. If they are not accepted within a reasonable time the offer may lapse, or be subject to some monetary adjustment should it later be accepted. The employer may stipulate in the invitation to the tenderers that the offer should remain open for a prescribed period of time.

Offers concerned with civil engineering projects are generally made on the basis of detailed terms and conditions. The parties to the contract will be bound by these conditions, as long as they know that such conditions were incorporated in the offer, even though they may never have read them or acquainted themselves with the details.

In many instances with civil engineering contracts, the offer must follow a stipulated procedure. Such procedures often incorporate delivery of the offer by a certain date and time, in writing, on a special form, and in a particular envelope, and stipulate that the offer must not be disclosed to a third party. Failure to comply with these procedures will result in the offer being rejected.

The acceptance

Once an agreeable offer has been made, there must be an acceptance of it before a contract can be established. The acceptance of the offer must be unconditional and it must be communicated to the person who made the offer. The unconditional terms of acceptance must correspond precisely with the terms of the offer. In practice, of course, the parties may choose to negotiate on the basis of the offer. For example, the promoter may require the project to be completed one month earlier and this may result in the tenderer revising the tender sum. A fresh offer (that is, a revised offer) made in this way is known as a counter-offer, and is subject to any revised conditions that may then be applied. An offer or acceptance is sometimes made on the basis of 'subject to contract'. In practice the courts tend to view such an expression as being of no legal effect. No binding contract will come into effect until the formal contract has been agreed.

Form

'Form' means some peculiar solemnity or procedure accompanying the expression of agreement. It is this formality which gives to the agreement its binding character. The formal contract in English law is the contract under seal – that is, one made by deed. The contract is executed – that is, it is made effective by being signed, sealed and delivered.

- Signature – Doubts have been expressed regarding the necessity for a signature. Some statutes make a signature a necessity.
- Sealing – Today this consists of affixing an adhesive wafer, or in the case of a corporate body an impression in the paper, and the party signs against this and acknowledges it as his seal.
- Delivery – This is not now necessary for legal effectiveness. As soon as the party acknowledges the document as the deed, it is immediately effective.
- Witnesses – These attest by signing the document. They are not usually a legal necessity.

Consideration

Another essential feature of a binding contract, other than a contract made under seal, is that the agreement must be supported by consideration. The most common forms of consideration are payment of money, provision of goods and the performance of work. Consideration has been judicially defined as 'some right, forbearance, detriment, loss or responsibility given, suffered or undertaken by the other in respect of the promise'. In civil engineering contracts the consideration of the contractor to carry out the works in accordance with the contract documents is matched by that of the employer to pay the price. The following rules concerning consideration should be adhered to:

- Every simple contract requires consideration to make it valid.
- The consideration must be worth something in the eyes of the law. The courts are not concerned whether the bargain is a good one, but simply that there is a bargain.
- Each party must get something in return for the promise, other than something already entitled to, otherwise there is no consideration.
- The consideration must not be such that it conflicts with the established law.
- The consideration must not relate to some event in the past.

Duress and undue influence

Duress is actual or threatened violence to, or restraint of the person of, a contracting party. If a contract is made under duress it is at once suspect, because consent has not been freely given to the bargain supposedly made. The contract is voidable at the option of the party concerned. Duress is a common law doctrine which relates entirely to the person and has no relation to that person's goods. As such it is a very limited doctrine and is one where cases are rare, especially in the construction industry.

Undue influence may occur where one party has a dominant position over the other. Undue influence will only exist, however, if the dominant party uses such influence unfairly.

Unenforceable contracts

Contracts may be described as void, or voidable, or unenforceable. A void contract creates no legal rights and cannot therefore be sued upon. It may occur because of a mistake as to the nature of the contract, or because it involves the performance of something illegal, that is prohibited by a statute. A void contract will also result due to the incapacity of the parties, as in the case of infants. Also corporations cannot make contracts beyond their stated powers; these are said to be *ultra vires*, i.e. 'beyond one's powers'.

A contract is said to be voidable when only one of the parties may take advantage. In cases involving misrepresentation, only the party who has been misled has the right to void the contract in one of the ways described above.

Unenforceable contracts are those that are valid, but owing to the neglect of the formalities involved, a party seeking to enforce it will be denied a remedy.

Mistake

The law recognises that, in some circumstances, although a contract has been formed, one or both of the parties are unable to enforce the agreement. The parties are at variance with one another and this precludes the possibility of any agreement. Mistake may be classified as follows:

- **Identity of subject matter** This occurs where one party intends to contract with regard to one thing and the other party with regard to another. The parties in this situation cannot be of the same mind and no contract is formed.
- **Identity of party** If the identity of either party enters into consideration, this will negate the contract where this was not known beforehand.
- **Basis of contract** If two parties enter into a contract on the basis that certain facts exist, and they do not, then the contract is void.

- **Expressing the contract** If a written contract fails to express the agreed intentions of the two parties, then it is not enforceable. Courts may, however, express the true intention of the parties and then enforce it as amended.

Misrepresentation

Misrepresentation consists in the making of an untrue statement which induces the other party to enter into a contract. The statement must relate to fact rather than an opinion. Furthermore, the injured party must have relied on the statement and it must have been a material cause of their entering into the contract. Where such a contract is voidable it may be renounced by the injured party, but until such time it is valid. Misrepresentation may be classified as:

- Innocent misrepresentation – where an untrue statement is made in the belief that it is true.
- Fraudulent misrepresentation – which consists of an untrue statement made with the knowledge that it is untrue or made recklessly without attempting to assess its validity.
- Negligent misrepresentation – is a statement made honestly, but without reasonable grounds for belief that it is true. It is really a special case of innocent misrepresentation, for, although the statement is made in the belief that it is true, insufficient care has been taken to check it.

Where misrepresentation occurs the injured party has several options:

- The injured party can affirm the contract, when it will then continue for both parties.
- The injured party can repudiate the contract and set up misrepresentation as a defence.
- An action can be brought for rescission and restitution. Rescission involves cancelling the contract and the restoration of the parties to the state that they were in before the contract was made. Restitution is the return of any money paid or transferred under the terms of the contract.
- The injured party can bring an action for damages. The claim for damages is only possible in circumstances of fraudulent misrepresentation.

Disclosure of information

When entering into a contract it is not always necessary to disclose all the facts that are available. A party may observe silence in regard to certain facts, even though it may know that such facts would influence the other party. This is summed up in the maxim *caveat emptor* – 'let the buyer beware'.

There are, however, circumstances where the non-disclosure of relevant information may affect the validity of the contract. This can occur where the relevant facts surrounding the contract are almost entirely within the knowledge of one of the parties, and the other has no means of discerning the facts. These contracts are said to be *uberrimae fidei* – 'of the utmost good faith'.

Privity of contract

A contract creates something special for the parties who enter into it. The common law rule of privity is that only the persons who are party to the contract can be affected by it. A contract can neither impose obligations nor confer rights upon others who are not privy to it. For example, a clause in the ICE Conditions of Contract which allows the employer to pay money direct to the subcontractor may be used by the employer. It cannot be enforced by the subcontractor, who is not a party to the main contract. The subcontractor may seek to persuade the employer to adopt this course of action but cannot enforce the employer to do so in a court of law.

Express and implied terms

The terms of contract can be classified as either express or implied. Those terms which are written into the contract documents or given orally by the parties are described as express terms. Those terms which were not mentioned by the parties at the time when they made the contract are implied terms, so long as they were in the minds of both parties.

Implied terms

Implied terms, although not expressly stated by the parties by words or conduct, are by law deemed to be part of the contract. Terms may be implied into contracts by custom, statute or the courts:

- Custom: In law this means an established practice or usage in a trade, locality, type of transaction or between parties. If two or more people enter into a contract against a common background of business, it is considered that they intend the trade usage of that business to prevail unless they expressly exclude it.
- Statute: There are many areas in the civil law where Parliament has interfered with the right of parties to regulate their own affairs. This interference mainly occurs where one party has used a dominant bargaining position to abuse this freedom. Thus in the sale of goods, the general principle *caveat emptor* (let the buyer beware) has been greatly modified, particularly in favour of the consumer by the provisions in the Sale of Goods Act 1979. In addition, there have been important changes in relation to exemption clauses brought about by the Unfair Contract Terms Act 1977.

- The courts: The court will imply a term into a contract under the doctrine of the implied term, if it was the presumed intention of the parties that there should have been a particular term, but they have omitted to state it expressly.

The courts will, where it becomes necessary, imply into civil engineering contracts a number of terms. Although the implied term is one which the parties probably never contemplated when making the contract, the courts justify this by saying that the implication is necessary in order to give business efficacy to the contract. This does not mean, however, that the courts will make a contract more workable or sensible. The courts, for example, will generally imply into a civil engineering contract the following terms:

- The contractor will be given possession of the site within a reasonable time, should nothing be stated in the contract documentation.
- The employer will not unreasonably prevent the contractor from completing the work.
- The contractor will carry out the work in a workmanlike manner.

Implied terms which have evolved from decided cases often act as precedents for future events. In the majority of the standard conditions of contract all of these matters are normally express terms, since the contracts themselves are very comprehensive and hope to cover every eventuality.

There are some notable terms which are not normally to be implied into civil engineering contracts, such as, for example, the practicability of the design. However, there will be express or implied terms that the work will comply with the appropriate statutes and regulations.

Express terms

An express term is a clear stipulation in the contract which the parties intend should be binding upon them. Traditionally, the common law has divided terms into two categories of conditions and warranties:

- Conditions: These are terms which go to the root of the contract, and for breach of which the remedies of repudiation or rescission of the contract and damages are allowed.
- Warranties: These are minor terms of the contract, for breach of which the only remedy is damages.

Limitations of actions

Generally speaking, litigation is a costly and time-consuming process which becomes more difficult as the time between the disputed events and the litigation increases. Also rights of action cannot be allowed to endure for ever. Parties to a contract must be made to prosecute their causes within a reasonable time. For this reason Parliament has enacted Limitation Acts which set a time limit on the commencement of litigation. The rules and procedures in respect of limitation of actions are contained in the Limitation Act 1980. The right to bring an action can be discharged in three ways:

- the parties to a contract might decide to discharge their rights
- through the judgement of a court
- through lapse of time

If an action is not commenced within a certain time the right to sue is extinguished. Actions in a simple contract (not under seal) and tort become statute barred after six years. In the case of contracts under seal, this period is extended to twelve years. The Act does permit certain extensions to these time limits in very special circumstances – for example, where a person may be unconscious as the result of an accident. If damage is suffered at a later date then this will not affect the limitation period, although an action for negligence may be pursued. The period of limitation can be renewed if the debtor acknowledges the claim or makes a part payment at some time during this period. A number of international conventions, particularly with respect to the law of carriage, lay down shorter limitation periods for action than those specified in the Act. For example, in carriage by air under the Warsaw Rules the period is two years.

The Unfair Contract Terms Act 1977

Contractual clauses designed to exonerate a party wholly or partly from liability for breaches of express or implied terms first appeared in the nineteenth century. The common law did not interfere but took the view that parties forming a contractual relationship were free to make a bargain within the limits of the law. This is still largely the rule today although the growth of large trading organisations has led to an increase in both excluding and restricting clauses to the severe detriment of other parties.

The efforts of the courts to mitigate the worst effects of objectionable exclusion clauses have been reinforced by the Unfair Contract Terms Act 1977. This Act restricts the extent to which liability can be avoided for breach of contract and negligence. The Act relates only to business liability and transactions between private individuals are not therefore covered. The reasonableness is further extended to situations where parties attempt to exclude liability for a fundamental breach, i.e. where performance is substantially different from that reasonably expected or there is no performance at all of the whole, or any part, of the contractual obligations. The following are subject to the Act's provisions:

- making liability or its enforcement subject to restrictive or onerous conditions
- excluding or restricting any right or remedy
- excluding or restricting rules of evidence or procedure
- evasion of the provisions of the Act by a secondary contract is prohibited.

Contra proferentem

Any ambiguity in a clause in a contract will be interpreted against the party who put it forward. It is a general rule of construction of any document that it will be interpreted *contra proferentem*, that is, against the person who prepared the document. As an exclusion clause is invariably drafted by the imposer of it, this is an extremely useful weapon against exclusion clauses. The effect of the rule is to give the party who proposed the ambiguous clause only the lesser of the clause protections possible. The one who draws up the contract has the choice of words and must choose them to show the intention clearly.

Agency

Agency is a special relationship whereby one person (the agent) agrees on behalf of another (the principal) to conclude a contract between the principal and a third party. Providing agents act only within the scope of the authority conferred upon them, those acts become those of the principal, and the principal must therefore accept the responsibility for them. The majority of contractual relationships involve some form of agency. For example, the engineer, in ordering extra work, is acting in the capacity of the employer's agent. The contractor may presume, unless anything is known to the contrary, that the carrying out of these extras will result in future payment by the client.

A contract of agency may be established by:

- express authority – authority that has been directly given to an agent by his principal
- implied authority – where, because a person is engaged in a particular capacity, others dealing with that person are entitled, perhaps because of trade custom, to infer that person has the necessary authority to contract within the limits usually associated with that capacity
- ratification – where a principal subsequently accepts an act done by the principal's agent, even where this exceeds the agent's authority. This becomes as effective as if the principal originally authorised it.

In contracts of agency a principal cannot delegate to agents powers which the principal does not already possess. The capacity of an agent is therefore determined by the capacity of the agent's principal.

Discharge of contracts

A contract is said to be discharged when the parties become released from their general contractual obligations. The discharge of a contract may be brought about in several ways.

Discharge by performance

In these circumstances the party has undertaken to do a certain task and nothing further remains to be done. In general, only the complete and exact performance of the contractual obligations can discharge the contract. In practice, where a contract has been substantially performed, payment can be made with an adjustment for the work that is incomplete.

A civil engineering contract is discharged by performance once the contractor has completed all the work, including the making good of defects under the terms of the contract. The engineer must have issued all the appropriate certificates and the employer paid the requisite sums. If, however, undisclosed defects occur beyond this period the employer can still sue for damages under the statute of limitations. The contract has not been properly performed if there are hidden defects.

Thus, A undertakes to sell to B 1,000 roof tiles. A will be discharged from the contract when the tiles have been delivered, and B when he has paid the price. A question sometimes arises whether performance by another party will discharge the contract. The general rule is that where personal qualifications are a factor of consideration, then that person must perform the contract. Where, however, personal considerations are unimportant it would not matter who supplied the goods.

Discharge under condition

Contracts consist of a large number of stipulations or terms. In many types of contract there are conditions which, although not expressed, are intended because the parties must have contracted with these conditions in mind. It will, however, be obvious that the terms and conditions of a civil engineering contract are not of equal importance. Some of the terms are fundamental to the contract, and are so essential that if they are broken the whole purpose of the contract is defeated.

Thus, if a contractor agreed to design and construct a civil engineering project for a specific use, and it is incapable of such a use, the employer would be able to reject the project and recover the costs from the contractor.

Civil engineering contracts frequently contain a number of terms forming a specification. The contractor agrees to construct the project in accordance with this specification. If the builder deviates from this in some small way then this will give rise to an action for damages and the contract will not be discharged.

It is, however, well established that, subject to an express or implied agreement to the contrary, a party who has received substantial benefit under a contract cannot repudiate it for breach of condition.

Discharge by renunciation

This is effected when one of the parties refuses to perform obligations. Thus, A employs an engineer to design and supervise a proposed civil engineering project. On completion of the design, A decides not to continue with the project. This is renunciation of the contract and the engineer can sue for fees on a *quantum meruit* basis – that is, for as much as has been earned.

Discharge by fresh agreement

A contract can be discharged by a fresh agreement being made between the parties, which is both subsequent to and independent of the original contract. Such a contract may, however, discharge the parties altogether. This is known as a rescission of the original contract. Where one party to a contract is released by a third party from undertaking obligations, then the discharge of the original contract is termed 'novation'. Any alteration to a contract made with the consent of the parties concerned has the effect of making a new contract.

Frustration

A contract formed between two parties will expressly or impliedly be subject to the condition that it will be capable of performance. If the contract becomes incapable of performance, the parties will be discharged from it. Impossibility of performance is usually called frustration of contract. It occurs whenever the law recognises that, without default of either party, a contractual obligation has become incapable of execution.

For example, the event causing the impossibility may be due to a natural catastrophe. A agrees to carry out a contract for the repair of a road surface, but prior to A's starting the work, the road subsides to such an extent that it disappears completely down the side of a hill. Impossibility may also occur because a government may introduce a law that makes the contract illegal. For example, A contracts to carry out some insulation work using asbestos. The government subsequently introduces legislation forbidding the use of this material in buildings, thus making the contract impossible to carry out. A contractor cannot, however, claim that a contract is frustrated if by deliberate actions a delay is caused to avoid completion. For example, a contract for the construction of a sea wall must be completed prior to winter weather setting in, otherwise practical performance may become impossible. The knowledge that this predictable event will occur, coupled with a deliberate delay, does not result in a frustration of the contract.

Examples of civil engineering contracts being frustrated are, however, extremely rare. It should be noted that hardship, inconvenience or loss do not decide whether an occurrence frustrates the contract, because these are accepted risks. The legal effect of frustration is that the contract is discharged and money prepaid in anticipation of performance should be returned. A party receiving benefit from a partly executed contract should reimburse the other party to the value of that benefit.

Determination of contract

There are provisions in all the common forms of contract that allow either party to determine (terminate the contract). There must, of course, be good reasons to support any party who decides to determine, otherwise a breach of contract can occur. All forms of contract include clauses that provide circumstances which allow the employer to terminate the contract with the contractor. In one sense they are fairly exceptional happenings, and so they should be, since they provide a final option to the employer. The decision to determine must be taken very carefully and reasonably. It must not be done vexatiously. When determination does occur it is frequently because the contractor is failing to take notice of relevant instructions from the engineer – perhaps to remedy an already existing breach of contract. Clauses provide for determination because of either a refusal or interference with payments due to the contractor. It may also occur where the works are suspended for an unreasonable length of time. Again the action to determine must be carefully taken, and is a last-resort decision on the part of the contractor. The latter takes this decision when it is felt it is no longer possible to continue, perhaps after a number of delays in payment, to work with the employer. Determination of contract by either party results in two losers and no winners. Although there are provisions in the contract for financial recompense to the aggrieved party, these rarely suffice in practice.

Assignment

It sometimes happens that one party to a contract wishes to dispose of its obligations under it, but the extent to which this is permitted is limited. The other party may have valid reasons why it prefers the obligations to be performed by the original contractor. The rule is that liabilities can only be assigned by novation. This is the formation of a new contract between the party who wishes performance and the new contractor, who is accepted as adequately qualified to perform as the original contractor. Liabilities can only be assigned by consent. By contrast, rights under a contract can usually be assigned without consent of the other party, except where the subject matter involves a personal service. However, even in those circumstances where the contract does not involve personal service, but specifically restricts the right to assign the contract or any interest in it to a third party, then the assignment of rights will not be permitted.

Remedies for breach of contract

A breach of contract occurs when one party fails to perform an obligation under the terms of the contract. For example, a breach of contract of the conditions of contract can occur if:

- The contractor refuses to obey an engineer's instruction.
- The employer fails to honour an engineer's certificate.
- The contractor refuses to hand over certain antiquities found on the site.
- The contractor fails to proceed regularly and diligently with the works.

Defective work is not necessarily in breach of contract, as long as the contractor rectifies it in accordance with an engineer's instruction. A breach will occur where the contractor either refuses to remove the defective work or ignores the remedial work required. A breach of contract may have two principal consequences:

- Damages
- If the breach is sufficiently serious, determination under the terms of the contract may result. The aggrieved party may decide to terminate the contract with the other, and also sue for damages, or alternatively take only one of these courses of action. Damages may include both the loss resulting from the breach, the loss flowing from the termination and the additional costs of completing the contract.

Damages

There are few exceptions to the rule that no one who is not a party to the contract can sue or be sued in respect of it. There may be other remedies, for instance in tort, but these are beyond the scope of this book.

The legal remedy for breach of contract is damages. These consist of the award of a sum of money to the injured party, designed to compensate for the loss sustained. The basis of the award of damages is by way of compensation. The damages awarded are made in an attempt to recompense the actual loss sustained by reason of the breach of contract. The injured party is to be placed, as far as is possible in monetary terms, in the same situation as if the contract had been performed. It should be clearly understood that not every breach automatically results in damages. In order for damages to be paid, the injured party must be able to prove that a loss resulted from the breach. Furthermore, the innocent party must take all reasonable steps to mitigate any loss. Damages may be classified as follows.

Nominal damages

Where a party can show a breach of contract, but cannot prove any sustained loss as a result, then nominal damages may be awarded. These comprise merely a small sum in recognition that a contractual right has been infringed.

Substantial damages

These damages represent the measure of loss sustained by the injured party. Despite their name they might be quite small.

Remoteness of damage

A breach of contract can, in some circumstances, create a chain of events resulting in considerable damage, and the question may arise whether the injured party or parties can claim for the whole of the damage sustained. In ordinary circumstances, the only damages that can be claimed are those which arise immediately because of the breach. Damage is not considered too remote if the parties at the time the contract was entered into considered the possibility that it could occur. If the contract specifies the extent of the liability, then no question of consequential damages arises. Defects liability clauses, which require the contractor to rectify defects, do not limit the extent of the remedial work, but may also require other work to be put right that has resulted from these defects.

Special damage

Damages resulting from special circumstances are recoverable if they flow from a breach of contract, and the special circumstances were known to both parties at the time of making the agreement.

Liquidated damages

Each of the standard forms of contract provides for payment of agreed damages by the contractor when completion of work is not within the stipulated time. These payments are known as liquidated damages, and their amount has to be stated in the contract documents, usually in an appendix to the contract. The sum stated should be a genuine estimate of the damage that the employer may suffer. If, however, the sum stated is excessive and bears no relation to the actual damage, then it may be regarded as a penalty. In these circumstances the courts will not enforce the amount stated in the contract but will assess the damage incurred on an unliquidated basis. Liquidated damages may be estimated on the basis of loss of profit in the case of commercial projects, but they are much more difficult to determine in the case of public works projects such as roads.

In some circumstances an employer may seek the occupation of a project even though it remains incomplete. This would normally deprive the employer of the right to enforce a claim for liquidated damages.

Unliquidated damages

When no liquidated damages have been detailed in the contract, the employer can still recover damages should the contractor fail to complete on time. The sum awarded in this case is that regarded as compensation for the loss actually sustained by the breach.

Other remedies

Specific performance

Specific performance was introduced by the courts for use in those cases where damages would not be an adequate remedy. In civil engineering contracts this remedy is only really available in exceptional circumstances. The courts will enforce a party to do what it has contracted to do, in preference to awarding damages to the aggrieved party. A decree of specific performance will not be granted where the court cannot effectively supervise or enforce the performance.

Injunction

Injunction is a remedy for the enforcement of a negative undertaking. A party is prohibited from carrying out a certain action. An injunction is often awarded by the courts as an effective remedy against nuisance.

Rescission (see Misrepresentation)

This is an equitable remedy, which endeavours to place the parties in the pre-contractual position by returning goods or money to the original owners. This means that the parties are no longer bound by the contract. This is granted at the discretion of the court but will not be awarded where:

- The injured party was aware of the misrepresentation and carried on with the contract.
- The parties cannot be returned to their original position.
- Another party has acquired an interest in the goods.
- The injured party waited too long before claiming this remedy.

Quantum meruit

A claim for damages is a claim for compensation for loss. Where, under the terms of the contract, one party undertakes its duty for the other, and the other party breaks the contract, the former can sue upon a *quantum meruit* basis – that is, to claim a reasonable price for the work carried out. It is payment that has been earned for work carried out. If the two parties cannot agree, the question of what sum is reasonable is decided by the courts. A claim on *a quantum meruit* basis is appropriate where there is an express agreement to pay a reasonable sum upon the completion of some work. In assessing a *quantum meruit* claim the parties may choose to use the various means available. For example, it may be based upon the costs of labour and materials plus profit, or the measurement of the work using reasonable rates.

CHAPTER 3

SETTLEMENT OF DISPUTES

Introduction

Construction contracts in the distant past consisted of a document a few pages long. They generally concluded with a handshake, but underlying such agreements were an essential set of values of competence, fairness and honesty. Today things are different. We have developed a complex and onerous set of conditions, for all parties, that attempt to cover every eventuality and, in so doing, create loopholes that the legal professions can feast upon. It has often been suggested that the only individuals to make any money in the construction industry are the lawyers. Precious time and resources are thus drawn away from the main purpose of getting the project built on time, to the right design and specification at an agreed price and standards and quality.

Many believe that construction contracts need to become less adversarial, more simply constructed and emphasise the positive needs of the project. Consultants and contractors should be allowed and encouraged to use their best endeavours and to work together as a team (see project partnering, Chapter 9) rather than each watching their individual back all the time. When a project ends up in a protracted dispute, it will fail to meet its original goals and expectations. In addition, employers will suffer from high legal fees, delayed completion and occupation and general dissatisfaction. The contractors' profits will diminish and they, too, will have additional legal fees. There are no winners under these circumstances.

The reasons why disputes arise

The construction industry is a risky business. It generally does not construct many prototypes, with each different project being individual in so many respects. Even apparently identical civil engineering projects that have been constructed on different sites create their own special circumstances, are subject to the vagaries of different site and weather conditions and use labour that may have different trade practices even from one site to another. The identical project constructed on an adjacent site by a different contractor will have different costs and different problems associated with its construction. Disputes are therefore likely to arise under the best circumstances, even where every possibility has been potentially eliminated. Some of the main areas for possible disputes occurring are given below.

General

- Adversarial nature of construction contracts
- Poor communication between the parties concerned
- Proliferation of forms of contract and warranties
- Fragmentation in the industry
- Tendering policies and procedures

Employers

- Poor briefing
- Changes and variation in requirements
- Changes to standard conditions of contract
- Interference in the contractual duties of the contract administrator
- Late payments

Consultants

- Design inadequacies
- Lack of appropriate competence and experience
- Late and incomplete information
- Lack of co-ordination
- Unclear delegation of responsibilities

Contractors

- Inadequate site management
- Poor planning and programming
- Poor standards of workmanship
- Disputes with subcontractors
- Delayed payments to subcontractors
- Co-ordination of subcontractors

Subcontractors

- Mismatch of subcontract conditions with main contract
- Failure to follow and adopt agreed procedures
- Poor standards of workmanship

Manufacturers and suppliers

- Failure to define performance or purpose
- Failure of performance

Issues for the resolution of disputes

The following are matters that need to be resolved in order to reduce the possibility of future disputes occurring:

- clarification of responsibilities
- need for single point responsibility contracts
- allocation of risk to the parties who are best able to control it
- further investigation of insurance-based alternatives
- need to develop and extend non-adversarial methods of dispute resolution
- partnership sourcing (contractors and consultants working in a consortium)
- quality management and quality assurance

The resolution of disputes includes the following, in ascending order of importance, outcomes and costs:

- contractual claims
- alternative dispute resolution
- arbitration
- litigation

Claims arising under a civil engineering contract fall into two categories: contractual claims and ex-gratia payments.

Contractual claims (see also Appendix 9)

These are claims that have a direct reference to conditions of contract. When the contract is signed by the two parties, the contractor and the employer, there is a formal agreement to carry out and complete the works in accordance with the information supplied through the drawings, specification and bills of quantities. Where the works constructed are of a different character or executed under different conditions then different costs will be involved. Some of these additional costs may be recouped under the terms of the contract, through, for example, remeasurement of the works. Other costs that the experienced contractor had not allowed for may need to be recovered under the heading of a contractual claim.

Ex-gratia and extra-contractual payments

These claims are not based upon the terms or conditions of contract. However, the carrying out of the works has resulted in loss and expense to the contractor. The contractor has completed the project on time to the required standards and conditions and sometimes a sympathetic employer may be prepared to make a discretionary payment to the contractor. Such payments are made out of grace and kindness. They are not common in the construction industry.

The ICE Conditions of Contract

The ICE Conditions of Contract seek to clarify the contractual relationship between the employer and the contractor. As far as possible ambiguities have been removed. If such forms or conditions of contract were not available, then the uncertainty between the two parties would be even greater. This could have the likely effect of increasing tender totals. Under the present conditions of contract, the contractual risks involved are shared between the employer and the contractor.

Notice of claims (clause 52 (4))

Under clause 52 (4) of the Conditions it is the contractor's responsibility to inform the engineer, within 28 days of an event that a claim may arise. The contractor must keep the necessary records in order to reasonably support any claim that may be subsequently made. Without necessarily accepting the employer's liability that a claim may exist, the engineer can instruct the contractor to keep and maintain such records. Where required the engineer is then able to inspect such records and have copies supplied where these are appropriate.

As soon as it is reasonably possible the contractor should provide the engineer with a written interim account providing full details of the particular claim and the basis upon which it is made. This should be amended and updated when necessary or when required. If the contractor fails to comply with this procedure, this might prejudice the investigation of the claim by the engineer and any subsequent payments by the employer to the contractor.

The contractor is entitled to have such amounts included in the payment of interim certificates under clause 60. However, a majority of claims are unlikely to be agreed until the end of the contract. In these circumstances the contractor is entitled to receive part of the claim included in an interim certificate where this can be substantiated by the engineer.

Contractual claims

Contractual claims arise where contractors assess that they are entitled to additional payments over and above those paid within the general terms and conditions of the contract. For example, the contractors may seek reimbursement for some alleged loss that has been suffered for reasons beyond their control. On many occasions the costs incurred lie where they fall and contractors will have recourse to recover them. Thus losses and delays arising from the intervention of third parties who are unconnected with the contract almost invariably fall with the contractor.

The fact that a loss has been sustained, without fault on the part of the loser, may merit sympathy, but does not in itself demand compensation.

Where a standard form of contract is used, many attempts may be made by contractors to invoke some of the compensatory provisions of the contract in order to secure further payment to cover the losses involved.

The details of such claims will be investigated by the measurement engineers or quantity surveyors and a report made to the engineer. The report should summarise the arguments involved and set out the possible financial effect of each claim. Considerable negotiation between the engineer and the contractor (or their representatives) takes place over such issues in an attempt to solve the problems and arrive at a reasonable and fair solution. This process is preferable to a lengthy and costly legal dispute. As in many issues in life, contractual claims are rarely one-sided and the fault of only one of the parties involved. However, where the claim cannot be resolved through negotiation, some form of legal proceedings may have to be initiated. This is usually instigated by the aggrieved party. Particular care therefore needs to be exercised in the conduct of the negotiations, since they may have an effect upon the outcome of any subsequent legal proceedings.

The contractor

Most civil engineering contractors have well organised systems for dealing with claims and the recovery of monies that are rightly due under the terms of the contract. They are likely to maintain good records of most important events, but particularly those where difficulties have occurred in the execution of the work. However, some of the difficulties may be due to the manner in which the contractor has sought to carry out the work and are thus the entire responsibility of the contractor.

Claims that are notified or submitted late will inevitably create problems in their agreement and payment. In these circumstances the engineer might not have the opportunity to check the details of the contractor's submission. Such occurrences will not be favourably looked upon by the engineer. The ICE Conditions suggest that where circumstances arise that may give rise to a potential claim the contractor should inform the engineer as soon as possible.

The contractor must prepare a report on why a particular aspect of the work has cost more than expected, substantiate this with appropriate calculations and support it with reference to engineer's instructions, drawings, specifications, letters, etc. The contractor must also be able to show that, as an experienced contractor, they could not have foreseen the difficulties now being explained. They will also need to satisfy themselves, and the engineer, that the work was executed in an efficient, effective and economic manner.

Claims are for additional payments that cannot be recouped in the normal way simply through measurement and valuation. They are based on the assumption that the works constructed differed considerably from the works for which the contractor originally submitted a tender. The differences may have changed the contractor's method of working and this, in turn, may have altered the costs involved. The rates inserted by the contractor in the bill of quantities are not now a true reflection of the work that has been executed.

Example

On a large earthmoving contract the quantities of excavation and its subsequent disposal off site to tips has increased in measured volume by 25 per cent. The actual quantities of excavated materials to be moved are larger due to bulking of the spoil. The contractor had arranged to use a variety of tips at different distances from the site. The contractor's tendering notes indicated that the tips nearer the site would be filled first. The disposal of the excavated material therefore includes two elements:

- the haul charges to the tip
- the costs of the tip

The costs of the new tips could result in lower charges for the additional materials, since not all tips will charge the same rate for disposal of spoil. In the bill of quantities the costs are based upon an average for all of the tips involved. Since the material has to be taken to a tip a greater distance away, this will increase the contractor's haulage costs.

Some of the factors that the contractor would have to consider include:

- The increase in the amount of excavated materials, on this scale, may also have other repercussions, such as an extension of time.
- Differences in the geological nature of the soil.
- The method adopted by the contractor for carrying out the works may be different from that originally envisaged by the contractor.
- The contractor may be involved in hiring additional plant at higher charges and employing workpeople at overtime rates.

The preparation of the claim includes two aspects:

- a report (the basis of the claim) outlining the reasons why additional payments should be made
- a calculation showing how the additional costs have been calculated

The contractor must, as a matter of good practice, always put in writing:

- applications for instructions, drawings, etc.
- application for the nomination of subcontractors
- notification of any claims under the contract in respect of variations or loss and expense connected with the progress of the works, and any delays
- confirmation of any oral instructions of the engineer

The contractor should also ensure that any certificates that are required under the terms of the contract are issued at the appropriate time. These may have some effect upon the validity or otherwise of a contractor's claim at a later stage.

The engineer

The engineer should recognise that whilst most civil engineering contractors desire to carry out the works to the complete satisfaction of the engineer and employer, their main reason for being in business is to make a financial profit from the project.

The engineer is very much in control of the project and can approve or disapprove of the contractor's methods of working. The engineer is able to issue instructions to the contractor under the terms of the contract, order additional works, approve or nominate subcontractors, etc. The engineer will also have had some input in appointing the firm of contractors before the contract was awarded. A contractor may have to request information from the engineer. This may be a positive step on the part of the contractor in alleviating a claim at some later stage. The engineer must respond to such requests within a reasonable time.

After receiving a contractual claim from the contractor the engineer should consider the following:

* Is the contractor's claim reasonable?
* What are the costs involved?
* What clauses in the Conditions of Contract are relevant?
* What basis is the contractor using to justify a claim?

It is easy to attempt to dismiss a contractor's claim out-of-hand. It is well recognised that claims generally, and contractors' are no exception, are often inflated on the assumption that the claim will be contested and then reduced. This is the basis of negotiation. The contractor is also unlikely to be faultless in the matter. The agreed amount is therefore always likely to be lower than that originally indicated or calculated by the contractor.

The engineer's first instinct may be to reject the claim, since it might appear to reflect upon those employed by the engineer. It may imply that the engineer has not been carrying out the duties under the contract correctly. Under these circumstances the employer might have some redress against the engineer for professional misconduct.

Where the matter cannot be resolved by the two parties concerned other ways of dealing with the problem are available to them.

Alternative dispute resolution (ADR)

Alternative dispute resolution is a non-adversarial technique which is aimed at resolving disputes without resorting to the traditional forms of either litigation or arbitration. The process was developed in the USA but has also been widely used elsewhere in the world. It is claimed to be fast, effective and less expensive. It is also less threatening and stressful. ADR offers the parties who are in dispute the opportunity to participate in a process that encourages them to solve their differences in the most amicable way possible. Table 3.1 illustrates a comparison between litigation, arbitration and ADR.

Table 3.1 Characteristics of resolution

Characteristics	Litigation	Arbitration	Alternative dispute resolution
Place/conduct of hearing	Public court; unilateral initiation; compulsory	Private (with few exceptions); bilateral initiation; voluntary (subject to statutory provisions)	Private; bilateral initiation; voluntary
Hearing	Formal; before a judge	Formal; conforming to rules of arbitration; before an arbitrator	Informal; before a neutral third party
Representation	Legal; lawyers influence settlement	Legal; lawyers influence settlement	Legal only if necessary; disputants negotiate settlement
Resolution/disposal	Imposed by a judge after adjudication; limited right of appeal	Award imposed by an arbitrator; limited right of appeal	Mutually accepted agreements; option of arbitration if dissatisfied
Outcome	Unsatisfactory; legal win or lose	Unsatisfactory; legal win or lose	Satisfactory; business relationship maintained
Time	Time-consuming	Can be time-consuming	Fast
Cost	Expensive; often uneconomic	Uneconomic	Economic

Source: based on Alternative Dispute Resolution (ADR) in A.A. Kwakye, *Construction*, CIOB Construction Papers No. 21, 1993

Before the commencement of an ADR negotiation, the parties who are in dispute should have a genuine desire to settle their differences without recourse to either litigation or arbitration. They must therefore be prepared to compromise some of their rights in order to achieve a settlement. Proceedings are non-binding until a mutually agreed settlement is achieved. Either party can therefore resort to arbitration or litigation if the ADR procedure fails.

Common forms of ADR

Conciliation (see also ICE conciliation procedure, p. 43)

This is a process where a neutral adviser listens to the disputed points of each party and then explains the views of one party to the other. An agreed solution may be found by encouraging each party to see the other's point of view. With this approach, the neutral adviser plays the passive role of a facilitator. Recommendations are not made by the adviser, any agreement is reached by the parties agreeing to settle their differences. Where an agreement is achieved, the neutral adviser will put this in writing for each of the parties to sign.

Mediation

In this scenario the neutral adviser listens to the representations from both parties and then helps them to agree upon an overall solution. An active role is played by the adviser by putting forward suggestions, encouraging discussions and persuading the parties to focus upon the key issues. Private discussions may be held with each party in order to explain the points concerned and to attempt to formulate a mutually acceptable solution to the problem. Where this is successful, the agreement is put in writing and signed by the parties concerned.

Executive tribunal

This is a more formal arrangement undertaken by a group comprising a neutral adviser together with representatives of the parties involved in the dispute. This group has the authority to settle the matters of dispute that created the conditions of ADR in the first place. At the hearing each party makes representations to the chairman and is also able to raise questions and seek points of clarification. Witnesses may be called upon to give evidence, although this is unusual. Each of the parties in dispute then attempts to settle the matter in private in order to achieve a negotiated settlement. Again the agreement is put in writing for endorsement by the parties concerned.

Whilst the above arrangements may appear to be separate approaches to ADR, in practice a combination of the different aspects may be used in order to solve the dispute.

ADR advisers

The neutral advisers are typically from a construction professional or a lawyer background. The latter are likely to have some knowledge of the construction industry and will probably have been involved in settling such disputes in the traditional way. Persons appointed as ADR advisers will need to be trusted by and acceptable to both parties, with known impartiality and fairness. They will seek to develop a process that suits the particular circumstances involved and to seek out the truth of the dispute through questioning and debate. ADR advisers will have good problem-solving abilities and communication skills and will be able to put each of the parties and the witnesses at ease in order to build their trust and confidence. They will, in addition, be able to appreciate all of the relevant issues involved and direct the parties towards the matters that are of crucial importance to the case.

Role of advisers in ADR

The role of advisers in ADR proceedings includes:

- explaining to the parties involved the procedures to be employed and the outcomes expected
- arranging the meetings, setting agendas and outlining the protocol to be used
- managing the process
- clarifying the issues involved and seeking conciliation rather than encouraging confrontation
- allowing each party to state and explain their case
- encouraging each party to see the other party's point of view
- preparing a report and assisting the parties to accept the agreement as binding

ICE Conciliation Procedure (1994)

The Institution of Civil Engineers, the Association of Consulting Engineers and the Federation of Civil Engineering Contractors (in future the Construction Employers' Confederation) have approved the above document and a permanent committee of the sponsoring authorities will keep it under review.

The main difference between conciliation and arbitration is that the outcome of conciliation is not imposed. It only becomes binding with the consent of each party. The conciliation procedure allows the parties to the dispute the freedom to explore ways of settling the dispute with the assistance of an independent impartial person, referred to as the conciliator. Conciliation with the ICE Conditions of Contract is essentially assisted negotiation.

The conciliation procedure includes 24 clauses together with a form signifying agreement of the parties to the contract. Some of the main points of this agreement include:

- Procedures to be used in the most conducive and efficient manner.
- Any party can send the conciliator a copy of their views.
- Confidentiality of the conciliation by the parties.
- Conciliator cannot be appointed arbitrator in subsequent proceedings.
- Conciliator fixes the date for the conciliation meeting.
- Additional parties may be invited to the proceedings.
- Additional claims or disputes may also be considered.

Arbitration

Arbitration is an alternative to legal action in the courts, in order to settle an unresolved dispute. No one is compelled to submit a dispute to arbitration unless they have agreed to do so within the terms of the contract. Once a person has agreed to this method of settling a disagreement they cannot then take legal proceedings until a settlement has been sought through arbitration. If they attempt to do so the courts will stay such proceedings. All the standard forms of contract used in the construction industry include an arbitration provision. It is, therefore, the procedure that is most commonly used for dealing with disputes that arise between the various parties concerned.

The essential features of a valid arbitration agreement are as follows:

- The parties must be capable of entering into a legally binding contract.
- The agreement should, whenever possible, be in writing.
- It must be signed by the parties concerned.
- It must state clearly those matters which will be submitted to arbitration and when the proceedings will be initiated.
- It must not contain anything that is illegal.

Arbitration Act 1979

This Act came into force on 4 April 1979 and extended the previous Act of 1950. The Act introduced a revised appeals procedure and reinforced the power of the arbitrator to proceed ex parte in some cases by adding the authority of the High Court. The Act consists of a number of provisions, some of which will apply to all arbitrations, while others depend on the arbitration agreement drawn up by the parties. The majority of arbitration agreements adopted by the construction industry accept the Act in its entirety.

Advantages of arbitration

- Arbitration is generally, though not always, less expensive than court proceedings.
- Arbitration is a more speedy process than an action at law. A delay of a year before a case comes before the courts is not uncommon.

- Arbitration hearings are usually held in private, thus avoiding any bad publicity that might be associated with a case in the courts.
- The time and place of the hearing can be arranged to suit the parties concerned. Court proceedings take their place in turn amongst the other cases and at the law courts concerned.
- Arbitrators are selected for their expert technical knowledge in the matter of the dispute. Judges do not generally have such knowledge.
- In cases of dispute which involve a construction site or existing structure, it can be insisted that the arbitrator visit the site concerned. Although a judge may decide upon a visit, this cannot be enforced by the parties concerned.

Disadvantages of arbitration

- The courts will generally always be able to offer a sound opinion on a point of law. The arbitrator may, of course, seek the opinion of the courts, but this could easily be overlooked, in which case a mistake could occur.
- An arbitrator does not have the power to bring into an arbitration a third party against their wishes. The courts are always able to do this.
- The arbitrator and the place of the meetings have to be paid for, whereas a judge and the courts are free.

Terminology

The following terminology is associated with arbitrations:

Arbitrator – The person to whom the dispute is referred for settlement. Arbitrators are often appointed by, for example, the President of the ICE, IStructE or CIArb. In practice they may be selected because of their expert technical knowledge of the subject matter in dispute.

Umpire – It may occasionally be preferable to appoint two arbitrators. In this event a third arbitrator, known as an umpire, is appointed to settle any dispute over which the two arbitrators cannot agree.

Reference – The actual hearing of the dispute by the arbitrator.

Award – The decision on the matter concerned made by the arbitrator.

Respondent – The equivalent of the defendant in a law court.

Claimant – The equivalent of the plaintiff.

Expert witness – A special type of witness who plays an important part in arbitrations. Ordinary witnesses must confine their evidence strictly to the statement of facts. Expert witnesses may, however, put forward their opinion based upon technical knowledge and practical experience. Prior to presenting evidence, they must show by experience and academic and professional qualifications that they can be recognised as an expert in the subject matter.

Appointment of the arbitrator

An arbitrator or umpire should be a disinterested person, who is quite independent of the parties involved in the proceedings and sufficiently qualified and experienced in the matter of the dispute. However, it is for the parties concerned to choose the arbitrator, and the courts will not generally interfere even where the person appointed is not really the most appropriate person to settle the dispute. Nevertheless, arbitrators may be disqualified if it can be shown that:

- They have a direct interest in the subject matter of the dispute (for example, where the decision may have direct repercussions on their own professional work).
- They may fail to do justice to the arbitration by showing a bias towards one of the parties concerned (for example, it could be argued that engineers might show favour to an employer since they are normally employed on this 'side' of the industry).

Each of the parties to the arbitration agreement must be satisfied that the arbitrator who is appointed will give an impartial judgement on the matter of the dispute. The remarks expressed by the arbitrator during the conduct of the case will probably reveal any favouring of one of the parties in preference to the other. This may lead to removal of the arbitrator by the courts on the grounds of misconduct.

Once the arbitrator has been appointed it will be necessary to establish the general matters relating to the dispute and to determine:

- the facilities available for inspecting the works
- whether the parties will be represented by counsel
- the matters the parties already agree upon
- the time and place of the proposed hearing.

Outline of the procedure

The pleadings

These are the formal documents which may be prepared by counsel or solicitor. The arbitrator will first require the claimant to set out the basis of the case. This will be included in the document, termed the 'point of claim', which will be served upon the respondent. The respondent then submits a reply in answer to the points of claim, termed the 'points of defence'. The respondent may also submit points of counterclaim, which will be served on the claimant at the same time. This may raise relevant matters that were not referred to in the points of claim. The claimant, in reply to the matters raised in the counterclaim, will submit points of reply and defence to the counterclaim.

The purpose of the above documents is very important, since they will make clear to the arbitrator the matters which are in dispute. Furthermore, the parties involved cannot stray beyond the scope of the pleadings without leave from the arbitrator.

Discovery

Once the pleadings have been completed, the precise issues which the arbitrator is to decide will be very clear. Every fact to be relied upon must be pleaded, but the manner in which it is to be proved need not be disclosed until the reference. 'Discovery' means the disclosure of all documents which are in the control of each party and which are in any way relevant to the issues of arbitration. Each party must allow the other to inspect and to take copies of all or any of the documents in their list, unless they can argue that it is privileged. The most important of these documents are the communications between a party and their solicitors for the purpose of obtaining legal advice. A party who refuses to allow inspection may be ordered to do so by the arbitrator.

In fixing the date and place of the hearing arbitrators have the sole discretion, subject to anything laid down in the arbitration agreement. They must, however, be seen to act in a reasonable manner. A refusal to attend the hearing by either party, after reasonable notice has been given, may empower the arbitrator to proceed without that party, i.e. ex parte.

The hearing

The procedure of the hearing follows the rules of evidence used in a normal court of law. The parties may or may not be represented by counsel or other representatives.

The claimant sets out the case, and calls each of the witnesses in turn. Once they have given their evidence they are cross-examined by the respondent (or counsel/representative where appropriate). The claimant is then allowed to ask these witnesses further questions on matters which have been raised by the cross-examination.

A similar procedure is then adopted by the respondent, who sets out the details of the counterclaim if this is necessary. Witnesses are examined and these in turn are cross-examined by the claimant. The claimant then replies to the respondent's defence and counterclaim, and presents a defence to the counterclaim. When this has been completed the respondent sums up the case in an address to the arbitrator, known as the respondent's closing speech. The claimant then has the right of reply or the last word in the case.

The trial is now ended and both parties await the publication of the award. The arbitrator may now decide to inspect the works, if this has not already been done, or re-examine certain parts of the project in more detail. The arbitrator will usually make the decision in private, and will set out the decision in the award. This is served on the parties after the appropriate charges have been met. It is enforceable in much the same way as a judgement debt, where the successful party can reclaim such costs as the arbitrator has awarded.

Evidence

To enable the arbitrator to carry out justice between the parties, the evidence must be carefully considered. This is submitted in turn by the claimant and the respondent. Evidence is the means by which the facts are proved. There are rules of evidence which the arbitrator must ensure are observed. These have been designed to determine four main problems.

- Who is to assume the burden of proving facts? Generally speaking, the person who sets forth a statement has the burden of determining its proof. The maxim, 'innocent until proved guilty' is appropriate in this context.
- What facts must be proved? A party must give proof of all material facts which were relied upon to establish the case, although there can be exceptions to this rule. For example, the parties may agree on formal 'admissions' in order to dispense with the necessity of proving facts which are not in dispute.
- What facts will be excluded from the cognisance of the court? In order to prevent a waste of time or to prevent certain facts from being put before juries, which might tend to lead them to unwarranted conclusions, English law permits proof of facts which are in issue and of facts which are relevant to the issue.
- How is proof to be effected? The law recognises three kinds of proof:
  - oral proof – statements made verbally by a witness in the witness box
  - documentary proof – contained in the documents that are available
  - real proof, which could include models or a visit to the site in order to view the subject matter

It is usual in arbitration for the evidence to be given on oath or affirmation. The giving of false evidence is perjury and is punishable accordingly by fine or imprisonment. The arbitrator has no right to call a witness, except with the consent of both parties. Witnesses will, however, usually answer favourably to the party by whom they are called, in order to further that party's case. They must not in general be asked leading questions, which attempt to put the answer in the mouth of the witness. For example, 'Did you notice that the scaffold was inadequately fixed?' is a leading question. It must be rephrased: 'Did you notice anything about the scaffolding?' Leading questions can, however, be used to the opposing party's witnesses. The arbitrator must never receive evidence from one party without the knowledge of the other. Where, for example, communications are received from one party the other party must immediately be informed.

An arbitrator must always refuse to admit evidence on the grounds that:

- The witness is incompetent, e.g. refuses to take the oath, too young, a lunatic.
- The evidence is irrelevant, e.g. evidence which, in the opinion of the arbitrator, has no real bearing upon the facts.

- The evidence is inadmissible, e.g. hearsay evidence, since it is not made under oath and is not capable of cross-examination. (There are, however, exceptions to this rule, such as statements made on behalf of one party against their own interests.) Where documents are used as evidence it is the arbitrator's responsibility to ensure their authenticity.

Where one party produces a document, this must be proved unless the other side accepts it as valid. Documents under seal must be stamped, and generally the original document must be produced wherever possible.

Stating a case

A question of law may arise during the proceedings, and the arbitrator may deal with this in one of three ways:

- decide the matter personally
- consult counsel or a solicitor
- state a case to the courts

Where the decision is to take the latter course of action, a statement is prepared, outlining clearly all the facts in order that the courts may decide a point of law. Once the court has given its decision, the arbitration proceedings can continue. The arbitrator must then proceed in accordance with the court's decision. Failure to do so results in misconduct on the part of the arbitrator. The arbitrator may voluntarily take this course of action or be required to do so by one of the parties.

The arbitrator may also state a case to the courts upon the completion of the arbitration. In this case the award will be based upon the alternative findings to be resolved by the courts. The decision that the courts reach will depend upon which alternative is to be followed. For example, the arbitrator may state that a particular sum should be paid by one of the parties to the other if the courts approve the arbitrator's view of the law. Where the courts do not confirm this opinion, the arbitrator will have also indicated the course of action to be taken.

The award

The arbitrator's award is the equivalent of the judgement of the courts. The award must be made within the terms of reference, otherwise it will be invalid and therefore unenforceable. The essentials of a valid award can be summarised as follows:

- It must be made within the prescribed time limit set by the parties.
- It should comply with any special agreements regarding its form or method of publication laid down in the arbitration agreement.
- It must be legal and capable of enforcement at law.
- It must cover all the matters which were referred to the proceedings.

- It must be final in that it settles all the disputes which were referred under the arbitration.
- It must be consistent and its meaning must be clear, not contradictory or ambiguous.
- It must be confined solely to the matters in question, and not touch on matters which are outside the scope of dispute.
- The award should generally be in writing, in order to overcome any problems of enforcing it in practice

Publication of the award

The usual practice is for the arbitrator to notify both the parties that the award is ready for collection upon the payment of the appropriate fees to the arbitrator. If the successful party pays the fees then they are able to sue the other party for that amount, assuming that the costs follow the event. If the award is defective or bad, or it can be shown that there have been irregularities in the proceedings, application may be made to the courts to have it referred back for reconsideration or set aside altogether.

Referring back the award

The grounds on which the court are likely to refer back an award are:

- where the arbitrator makes a mistake so that it does not express the true intentions
- where it can be shown that the arbitrator has misconducted the proceedings, for example, in hearing the evidence of one of the parties in the absence of the other
- where new evidence, which was not known at the time of the hearing, comes to light and as such will affect the arbitrator's award.

The application to refer back an award should be made within 6 weeks of the award being published. The courts have the full discretion regarding the costs of an abortive arbitration. There is, of course, the right of appeal against the court's decision. The arbitrator's duty in dealing with the referral will depend largely upon the order of the court. As a rule, fresh evidence will not be heard unless new evidence has come to light. The amended award should normally be made within 3 months of the date of the court's order.

Setting aside the award

When the courts set aside an award it becomes null and void. The situations where the courts will do this – similar to those for referring back an award, but much more serious are:

- where the award is void, for example, if the arbitrator directs an illegal action

- the discovery of evidence that was not available at the time the arbitration proceedings were held
- where the arbitrator has made an error on some point of law
- misconduct on the part of the arbitrator by permitting irregularities in the proceedings
- where the award has been obtained improperly, for example, through fraud or bribery
- where the essentials of a valid award are lacking, for example, the award is inconsistent or impossible of performance

Misconduct by the arbitrator

Arbitrators must carry out their duties in a professional manner. Where they are guilty of misconduct, the award can be referred back by the courts, and in a serious case the effects of setting aside an award are to make it null and void. Misconduct may be classified as 'actual' or 'technical'.

Actual misconduct would occur where arbitrators have been inspired in their decisions by some corrupt or improper motive or have shown bias to one of the parties involved.

Technical misconduct is when some irregularities in the proceedings occur, and may include:

- the hearing of evidence of one party in the absence of the other
- the examination of witnesses in the absence of both parties
- the refusal to state a case when requested to do so by one of the parties
- exceeding jurisdiction beyond the terms of the reference
- failing to give adequate notice of the time of the proceedings
- delegating authority
- an error of law

Although an arbitrator may be removed because of misconduct, this will not terminate the arbitration proceedings in favour of, say, litigation. The courts, on the application of any party to the arbitration agreement, may appoint another person to act as arbitrator. The courts may also remove an arbitrator who has failed to commence the proceedings within a reasonable time or has delayed the publishing of the award.

If the arbitrator dies during the proceedings, the arbitration will not be revoked. The parties must agree upon a successor.

The parties may at any time, by mutual agreement, decide to terminate the arbitrator's appointment. Where duties have been commenced, there will be entitlement to some remuneration.

Costs

An important part of the arbitrator's award will be the directions regarding the payment of costs. These costs, which include the arbitrator's fees, can sometimes exceed the sum which is involved in the dispute. The arbitrator must exercise discretion regarding costs, but should follow the principles adopted by the courts.

The agreement generally provides for the 'costs to follow the event', which means that the loser will pay. Where an arbitration involves several issues, and the claimant succeeds on some but fails on others, the costs of the arbitration will be apportioned accordingly. If the award fails to deal with the matter of costs, then any party to the reference may apply to the arbitrator for an order directing by whom or to whom the costs shall be paid. This must be done within 14 days of the publication of the award. A provision in an arbitration agreement that a party shall bear their own costs or any part of them is void.

The entire costs of the reference, which will include not only the costs of the hearing but also the costs incurred in respect of preliminary meetings and matters of preparation, are subject to taxation by the courts. This involves the investigation of bills of costs with the objective of reducing excessive amounts and removing improper items. For example, if a party has instructed an expensive counsel which the issues involved do not justify, then the fees paid will be substantially reduced. If witnesses have been placed in expensive hotels, then this may be struck from the claim as an unnecessary expense. Such items are known as 'solicitor and client charges' and have usually to be borne by the successful party.

Litigation

Litigation is a dispute procedure which takes place in the courts. It involves third parties who are trained in the law, usually barristers, and a judge who is appointed by the courts. This method of solving disputes is often expensive and can be a very lengthy process before the matter is resolved. The process is frequently extended to higher courts, involving additional expense and time (see Chapter 1). Also, since a case needs to be properly prepared prior to the trial, a considerable amount of time can elapse between the commencement of the proceedings and the trial, as noted above.

A typical action is started by the issuing of a writ. This places the matter on the official record. A copy of the writ must be served on the defendant, either by delivering it personally or by other means such as through the offices of a solicitor. The general rule is that the defendants must be made aware of the proceedings against them. The speed of a hearing in most cases depends upon the following:

- availability of competent legal advisers to handle the case, i.e. its preparation and presentation
- expeditious preparation of the case by the parties concerned
- availability of courts and judges to hear the case

The amount of money involved in the case will determine whether it is heard in the County Court or High Court. Where the matter is largely of a technical nature the case may be referred in the first instance to the Official Referee's Court. An Official Referee is a circuit judge whose court is used to hearing commercial cases, and hence handles most of the commercial and construction disputes.

Under these circumstances a full hearing does not normally take place, but points of principle are established. The outcome of this hearing will determine whether the case then proceeds towards a full trial. Under some circumstances, the plaintiff may apply to the court for a judgement on the claim (or the defendant for a judgement on the counterclaim), on the ground that there is no sufficient defence. Provided that the court is satisfied that the defendant (or plaintiff) has no defence that warrants a full trial of the issues involved, judgement will be given, together with the costs involved.

Every fact in a dispute that is necessary to establish a claim must be proved to the judge by admissible evidence, whether oral, documentary or of other kind. Oral evidence must normally be given from memory by a person who heard or saw what took place. Hearsay evidence is not normally permissible.

In a civil action the facts in the dispute must be proved on a balance of probabilities. This is unlike a criminal case, where proof beyond reasonable doubt is required. The burden of proof usually lies upon the party asserting the fact.

Settlement of disputes (clause 66)

Where disputes arise under the ICE Conditions of Contract, the contract identifies how the dispute should first be settled. This is to avoid the contractor rushing off to the courts at the first hint of a problem. The Conditions of Contract cover possible disputes that might arise:

- during the progress of the works
- after completion
- before or after determination
- due to abandonment
- due to breach of contract

The dispute may arise because of the engineer's action in respect of a:

- decision
- opinion
- instruction
- direction
- certificate
- valuation

Notice of dispute

An official dispute arises when one party serves a notice on the engineer. This notice will state the nature of the dispute. The notice should be seen as a last resort measure, where all other opportunities for the avoidance or settlement of the dispute have been taken within the terms of the contract. Before the notice is served, the engineer should be allowed reasonable time to take the steps that are required.

Engineer's decision

Every dispute that arises must be dealt with by the engineer. The engineer's decision must be in writing. It must also be given to both the employer and the contractor within three calendar months.

Effect on contractor and employer

Unless the contract has already been determined or abandoned, the contractor should continue with the construction of the works. This should be done with all due diligence. The decision of the engineer is final and binding (at this stage), unless and until the recommendations of a conciliator are accepted by both parties or the decision of the engineer is revised by the publication of an arbitrator's award.

Conciliation

In relation to any dispute notified as above then after the engineer has given a decision or the time for giving a decision has expired, either party may give notice in writing that the dispute should be considered under the Institution of Civil Engineers' Conciliation Procedure (1988). The recommendation of the conciliator is deemed to be accepted by the parties unless a written notice is served within one calendar month of its receipt.

Arbitration

Prior to the issue of the certificate of substantial completion of the whole of the works if:

- the employer is dissatisfied with any decision of the engineer regarding the settlement of a dispute
- the contractor is dissatisfied with any decision of the engineer regarding the settlement of a dispute
- the engineer has failed to give a decision within a period of one calendar month after the service of a notice of dispute
- the employer is dissatisfied with any recommendation of a conciliator who has been appointed

- the employer is dissatisfied with any recommendation of a conciliator who has been appointed

then either the employer or the contractor may refer the dispute to arbitration

- within three calendar months of receiving notice of such decision, or
- within three calendar months of the expiry of the above period of one month, or
- within one calendar month of the receipt of the conciliator's recommendation.

The parties agree upon the nomination of an arbitrator.

Where the Certificate of Substantial Completion of the whole of the works has already been issued, the above provisions will continue to apply. The one calendar month period referred to above in this case will be three calendar months.

President or Vice-President to act

The appointment of an arbitrator should be made within one calendar month. If this is not done, then either of the parties can apply to the president of the Institution of Civil Engineers to appoint an arbitrator.

If the appointed arbitrator

- declines the appointment
- is removed by a competent court
- is incapable of acting
- dies

then if a new appointment has not been made within one calendar month the president of the Institution of Civil Engineers can be invited to appoint another arbitrator. Where the president is not able to exercise these functions then the vice-president will act on behalf of the president.

Arbitration – Procedures and powers

Any reference to arbitration in the Conditions of Contract is deemed to refer to the provisions of the Arbitration Acts 1950 and 1979 or their subsequent statutory revisions or amendments. The reference is to be conducted in accordance with the Institution of Civil Engineers' Arbitration Procedure (1983) or any revision or amendment.

- Under the Conditions of Contract an arbitrator has the power to open up, review and revise any of the engineer's
  - decisions
  - opinions
  - instructions
  - directions

- certificates
- valuations
- Neither party is limited in the proceedings before an arbitrator to the evidence or arguments put before the engineer.
- The award of the arbitrator is binding upon both parties unless overturned by the High Court, on appeal.
- Arbitration can be carried out at any time, it does not have to wait until the completion of the works.

Engineer as witness

No decision given by the engineer in accordance with the foregoing provisions will disqualify the engineer from being called as a witness and giving evidence before the appointed arbitrator.

PART TWO

PROCUREMENT

CHAPTER 4

FORMS AND CONDITIONS OF CONTRACT

Introduction

The United Kingdom is fortunate to have the benefit of a wide variety of published institutional standard forms or conditions of contract available for use in the civil engineering and building industries. However, a number of government-sponsored reports have also highlighted that this has major disadvantages. They identify the following disadvantages of proliferation:

- duplication of effort
- wasteful use of resources
- repetition
- inconsistency in practice
- vested interests

Whilst it would appear to represent a common-sense view to develop a single form or conditions of contract for the entire construction industry it is unlikely that this will be developed in the foreseeable future. There are too many vested interests and even amongst a single interest group, most admit the need for separate forms for major and minor works. There is also the international community to consider. The ICE (Institution of Civil Engineers) form, for example, has more in common with international forms such as FIDIC (Fédération Internationale des Ingénieurs-Conseils) than it has with forms that are used in the building industry such as JCT (Joint Contracts Tribunal).

It has been suggested that many of the different conditions of contract probably help to fuel the adversarial nature of the construction industry in which they are applied. It can also be argued that writing and interpreting the clauses of the various forms alone represents an industry in itself.

The widespread use of 'standard' conditions of contract within the construction industry is partly accounted for by the practical impossibility of writing a set of new contract conditions for every project, even if this was in any way desirable.

The construction of civil engineering and building projects represents a major investment for any employer. In some cases this will constitute the single highest purchase ever made in an organisation. This emphasises the importance of choosing the appropriate conditions of contract. For those employers who undertake construction projects as a regular part of their activities, the correct choice of conditions of contract is even more vital. The application of the principles and practices involved may be seen as precedents in the administration of their other contracts.

Since large sums of money are likely to be involved in these activities, it is crucial that the contractual arrangements should always be formal, legal and robust from the outset of the project. Where an employer allows consultants or contractors to begin their work on an informal basis, the employer's bargaining position is thereafter weakened or even eliminated. Under these circumstances, a worst case scenario for employers is that they may expect to spend years, often at substantial legal expense, in arguing over the precise nature of the contractual arrangements which should have been clarified from the start of the project.

Types of contract envisaged

The typical civil engineering or building contract refers to the contractual arrangement between the employer and contractor. The large part of this book has been written with this in mind. However, it should be remembered that with the construction of most civil engineering and building projects, contracts also need to be formed between other individuals or firms. These include, in addition to the main contract:

- engagement of the different consultants, e.g. engineers, architects, surveyors
- appointment of nominated subcontractors
- appointment of domestic subcontractors

In addition, the different contractual arrangements also require collateral warranties and performance bonds and, in some cases, personal and parent company guarantees.

Employer and contractor

As long ago as 1964, the Banwell Report recommended the use of a single form of contract for the whole of the construction industry. This was deemed to be both desirable and practicable. More recently the Latham Report, published in 1994, has reiterated these comments. Chapter 9 considers some of the issues raised in these reports in more detail. Unfortunately, since 1964, this apparently good suggestion has been thwarted, with just the opposite taking place. Since that time there has been a plethora of different forms and conditions of contract designed to suit the individual interests of particular employers and changes in the way that different sorts of construction work are now often procured. The different forms and conditions of contract reflect the vested interests of different parties and institutions, who, for a variety of reasons, wish to retain their separate identity whilst at the same time incorporating good practices. The industry is also fragmented at every level from education to practice. This is not just the separation between civil engineering and building works, but there are sub-categories within these main divisions.

The better forms and conditions of contract have incorporated the views of the different interested groups within the construction industry, such as employers and contractors. This represents a much fairer way of doing business and should result in less confrontation and conflict. Such forms or conditions have been prepared as a joint effort between the various parties involved. The ICE forms of Contract, for example, include a wide representation from different organisations in the construction industry.

The widespread use of different forms of contract is, of course, exacerbated where international projects are concerned. Not only are additional forms required, but the procurement methods can also be considerably different from those used in the United Kingdom. The law of the country concerned may also be different. It should be understood at this point that aspects of the laws of England and Scotland themselves continue to remain at variance! (See ICE Conditions of Contract clause 67.) Where there has been a British influence at an international level, e.g. amongst the Commonwealth countries, forms and conditions of contract resemble those used in the United Kingdom.

As one would expect, there are common threads that run through all the forms of contract regarding the employer's requirements and the contractor's obligations, payments, variations, quality, time, etc. Nevertheless, although the general layout and content of the various forms may appear somewhat similar, the details of application and procedure often vary considerably. It should be noted that the interpretation of the individual clauses will also differ. In some cases these have been clarified through the application of case law resulting from previous differences of opinion being settled in a court of law. The principles of the case law, however, usually only apply to the form or conditions of contract in question and cannot therefore be applied universally across all of the different forms.

The selection of a particular form or conditions of contract is dependent upon a number of different factors. These include:

- **Type of work to be performed:** civil engineering, process plant engineering, building
- **Size of project:** some forms and conditions are available for major and minor works and even those of an in-between nature
- **Status of designer:** civil engineers will usually select the ICE or NEC; architects are more likely to select a choice from JCT or ACA
- **Public or private sector:** different forms and conditions of contract are available for use by private employers and local and central government.

Large industrial corporations may in addition have developed their own forms and conditions of contract

- Procurement method to be used, e.g. traditional, design and construct, management, etc
- Methods used for calculating the costs, e.g. measurement, cost reimbursement, lump sum.

A major advantage of using a standard document is that those who use it regularly become familiar with its contents and able to apply it more easily and more consistently in practice. Individuals become aware of the strengths and weaknesses of a particular form and conditions of contract. They are able to identify the potential areas where disputes may arise, and take corrective action where this is possible. They are also able to identify the form's or conditions' suitability for projects with which they are concerned. The range of forms adopted by consultants is often more limited than those faced by contractors.

Main contract forms

The Institution of Civil Engineers (ICE) Conditions of Contract remains the most popular form of contract in the United Kingdom for works of civil engineering construction. In addition to the main form there is also a minor works form and a form for design and construction. The ICE form has many similarities with the FIDIC form that is used at an international level.

On building contracts a 'family' of forms of contract is available through the Joint Contracts Tribunal (JCT). JCT80 remains the most popular form of contract for building contracts in the United Kingdom. This has commonly been referred to as the Standard Form of Building Contract (SFBC), first published in 1980. Due partially to the excessive complexity of administering aspects of JCT 80, particularly in respect of nominated subcontractors, the Joint Contracts Tribunal introduced an Intermediate Form of Contract (IFC 84) in 1984. The reason for introducing this form was to provide contract conditions that were more appropriate for use on 'medium-sized' building projects. In practice, this form has received a more widespread use, often on the sort and size of projects that should have adopted JCT 80 as the preferred form of contract. JCT also publishes a minor works form of contract.

Civil Engineering

ICE Conditions of Contract and Form of Tender, Agreement and Bond for Use in Connection with Works of Civil Engineering Construction (ICE form) (sixth edition)

This is published jointly by the Institution of Civil Engineers (ICE), the Association of Consulting Engineers (ACE) and the Federation of Civil Engineering Contractors (FCEC) (to become the Construction Employer's Confederation).

The sixth edition was published in 1991. It was reprinted with amendments in November 1995.

The Conditions of Contract also includes:

- form of tender
- appendix to form of tender
- form of agreement
- form of bond

In addition, the Conditions of Contract includes for inclusion at the discretion of the user:

- contract price fluctuations clause
- contract price fluctuations fabricated structural steelwork clause
- contract price fluctuations civil engineering work and fabricated structural steelwork clause

Published separately to the Conditions of Contract is a set of guidance notes that have been prepared by the Conditions of Contract Standing Joint Committee (CCSJC) on the sixth edition. One of the main purposes of this committee is to consider amendments and to draft new standard forms. This committee is therefore in a good position to provide advice and guidance on the use of the form. The guidance notes do not purport to provide legal interpretation.

ICE Conditions of Contract and Form of Tender, Agreement and Bond for Use in Connection with Works of Civil Engineering Construction (ICE form) (fifth edition)

Although the sixth edition of the ICE Conditions of Contract was published in 1991, copies of the fifth edition are still available. There is still a significant demand from practitioners for this contract. The Institution of Civil Engineers has prepared a comparison between its fifth and sixth editions of the Conditions of Contract. This compares clause by clause the differences and similarities between the two editions.

The first edition of the ICE Conditions of Contract was published in 1945 by the Institution of Civil Engineers and the Federation of Civil Engineering Contractors.

ICE Conditions of Contract for Design and Construct

These conditions of contract have been developed by the CCSJC from the ICE Conditions of Contract, sixth edition. They were formally introduced in 1992. The form has been designed to meet a widely felt need in the construction industry for a form of contract for use over the whole range of design and construct situations. Notwithstanding its origins, this contract is not a version of the ICE Conditions sixth edition with selected modifications. It is a different form in its own right.

Guidance notes have also been provided explaining good practices and differences between the ICE Design and Construct Contract and other ICE contracts. The guidance notes have been prepared by the CCSJC.

ICE Conditions of Contract for minor works (second edition)

This contract has been prepared under the auspices of the three engineering bodies, ICE, ACE and FCEC. It was first introduced in 1988, with the current second edition dated 1995. It is intended for use on small works projects, originally of up to about £100,000 but the limit has recently been raised to £250,000. In practice it is used on much larger projects, particularly where these are straightforward in design and concept.

One of the reasons for updating this contract was to bring it into alignment with the sixth edition of the ICE Conditions. A set of guidance notes is provided to assist in its understanding and implementation.

Institution of Civil Engineers Conditions of Contract for Ground Investigation

These were first published in 1983 for use on ground investigation contracts.

Conditions of Contract (International) for Works of Civil Engineering Construction

This form is prepared by FIDIC (Fédération Internationale des Ingénieurs-Conseils – the International Federation of Consulting Engineers). The fourth edition was published in 1987. It is approved by several other organisations representing the construction interests of various other countries. It is well known, internationally recognised and accepted and adequately reflects the interests of the parties concerned. Whilst its primary purpose and use is for international contracts, it can be used on the domestic scene. This contract is printed in several languages.

New Engineering Contract (NEC)

The NEC provides a totally new approach in how engineering and construction contracts are structured and managed. Its aim is to provide a variety of contract strategies which can be adopted as necessary to suit different project, employer and contractor requirements. At the same time the NEC aims to provide a stimulus whereby all parties strive towards completion of a project without the disputes and adversarial approach inherent in some other contract strategies. It has been suggested that this contract could be renamed the New Construction Contract (NCC) in the future, in a bid to gain wider application and acceptance on building as well as civil engineering works.

To date however, it has mostly been used abroad, particularly in South Africa (85 per cent). Its use in the UK has been less than 5 per cent of its worldwide usage (based on about 700 contracts). Those who have used the NEC have also generated widely differing responses.

This form has generated a Contract Users' Group in order to bring together users and potential users of the NEC, professional institutions and others to exchange information and experience on the use of the NEC family of contracts. The aims of the Group are:

- to offer guidance on the application of the NEC
- to disseminate information about developments and applications
- to bring users together to share their experiences and knowledge.

The NEC is published in ten documents (see below), available either separately or combined in a slipcase. A professional service contract and an adjudicator's contract together with guidance notes is also available as a set.

1. New Engineering Contract
2. New Engineering Subcontract
3. Flow Charts
4. Option A: Priced Contract with Activity Schedule
5. Option B: Priced Contract with Bill of Quantities
6. Option C: Target Contract with Activity Schedule
7. Option D: Target Contract with Bill of Quantities
8. Option E: Cost Reimbursable Contract
9. Option F: Management Contract
10. Guidance Notes

General Conditions of Government Contracts for Civil Engineering And Building Works (GC/Works/1)

This form is published by HMSO. Most of the forms discussed above are prepared by joint bodies representing the various interested parties. This form is prepared solely by the employing agency, notably the Department of the Environment. It is used almost exclusively on projects authorised by central government departments and is not really suitable for other kinds of contracts. GC/Works/1 is used on major projects. The supervising officer (SO) controls the works. On civil engineering projects the SO is likely to be a chartered civil engineer.

There are in addition other related forms for both mechanical and electrical services in buildings.

General Conditions of Government Contracts for Civil Engineering And Building Works (GC/Works/2)

This is a similar form to GC/Works/1, but for use on minor works up to a value of about £200,000. Like many of the minor works forms it assumes the use of a drawing and specification only and a bill of quantities is not required.

General Conditions of Contract for Civil Engineering and Building Works (PSA/1)

This is a standard form of contract, for use in both the public and private sectors throughout the United Kingdom. It was developed from GC/Works/1. It is intended for use with bills of quantities, where all or most of the quantities are firm and not subject to remeasurement, giving a lump sum contract subject to adjustment for variations ordered.

Building

Joint Contracts Tribunal (JCT) forms

The JCT is widely represented from within the construction industry. Its constituent bodies are:

- Royal Institute of British Architects
- Building Employers' Confederation (being renamed as the Construction Employers' Confederation)
- Royal Institution of Chartered Surveyors
- Association of County Councils
- Association of Metropolitan Authorities
- Association of District Councils
- Confederation of Associations of Specialist Engineering Contractors
- Federation of Associations of Specialists and Subcontractors
- Association of Consulting Engineers
- British Property Federation
- Scottish Building Contract Committee

The following are the different forms of contract available from JCT. The date when the form was first introduced is shown alongside its abbreviated letters, e.g. JCT 80. The second date provided lists the latest edition of the form. In some cases additional amendments may have been issued, that have yet to be fully incorporated.

1. Joint Contracts Tribunal Standard Form of Building Contract (JCT 80) (1995): This is known colloquially as JCT 80. There are separate editions for use with either local authorities or the private sector. There are also different versions available for each: with quantities, without quantities or with approximate quantities. The six different versions are, however, very similar in their content throughout. Since its introduction in 1980, there have been fifteen amendments. An amendment TC/94 was also introduced in April 1994 for terrorism cover.

There are also a number of supplements to cover fluctuations (three alternatives), sectional completion and contractor's designed portion. The latter should not be confused with design and build, which is an entirely different concept, with its own particular JCT form (CD 81).

There are currently about twenty-eight JCT 80 Standard Form of Building Contract practice notes. These are not finally authoritative. They often deal with a practical application of the form, and contain material which will be included in future revisions to the form. Such notes, although they express only an opinion in legal terms, are nevertheless considered to be based upon expert legal advice. They will not, however, have the same binding effect as the decisions from courts of law.

There is also a series of guidance notes that have been prepared for use with JCT contracts on subjects such as:

- Joint Venture Tendering for Contracts in the UK (1985)
- Performance Bonds (1986)
- Collateral Warranties (1992)
- Alternative Dispute Resolution (1993)

2. JCT Intermediate Form of Contract (IFC 84) (1992): This form was introduced in the mid-1980s in response to the over-complexity of some of the provisions and procedures of JCT 80.

3. JCT Agreement for Minor Building Works (MW 80) (1994): This was introduced at the same time as JCT 80. It contains similar provisions to JCT 80, but in a much more simplified format. It is intended for use on minor building works schemes and small projects to be carried out on a lump sum and where an architect or supervising officer has been appointed on behalf of the employer.

4. JCT Standard Form of Building Contract with Contractor's Design (CD 81) (1991): This is used on projects of a design and build nature.

5. JCT Management Works Contract (MW 87) (1987): This is for use on management contracts.

6. JCT Prime Cost Contract (PCC 92) (1992): This form is used on cost reimbursement contracts.

7. JCT Standard Form of Measured Term Contract (MTC 89) (1992): This form is designed for use by employers in both the public and private sectors who need to undertake regular maintenance, building improvements and minor works programmes.

8. JCT Standard Form of Building Works of a Jobbing Character (JA 90) (1990): This contract is designed for use by local authorities and other employers who place a number of small jobbing contracts (up to £10,000 at 1990 prices) with various contractors and who are experienced in ordering jobbing work and dealing with contractors' accounts.

Subcontract forms

Civil engineering

FCEC (Federation of Civil Engineering Contractors) form of subcontract

The sixth edition of this form was prepared in conjunction with the ICE form. Presumably under the new Construction Employers' Confederation this form will be retained. It is likely that it will be renamed at its next revision. Separate editions of the form exist to coincide with the fifth and sixth edition of the main contract conditions.

Form of subcontract (international)

This form has been prepared for use with the FIDIC main form and conditions of contract.

The engineering and construction subcontract

This forms a part of the New Engineering Contract, described above.

Building

Under JCT 80 there are:

- nominated subcontracts for JCT 80 (NSC) (1992)
- nominated Supplier's Form of Tender
- domestic form of subcontract (DOM) (prepared by (BEC)
- named subcontractors for use with IFC 84
- standard form of subcontract for nominated subcontractors
- labour-only subcontract form

ICE legal notes

The Institution of Civil Engineers has prepared a number of legal note booklets that represent a series of guidance notes. These bring together and clarify important or contentious issues that affect the civil engineer. The information combines points of law with principles derived from the Rules of Professional Conduct. The booklets are easy to follow and understand, the information is free from legal jargon and offers direct guidance for action. The following are titles in the current series:

- *Direct professional access to counsel*
- *Collateral warranties*
- *Reviewing the work of another engineer*
- *Liability for latent damage*

- *Guidance memorandum for expert witnesses and their clients*
- *Engineers and auditors*

Appointment of consultants

Whilst 'the construction contract' is between the employer (employer or promoter) and the contractor, the various consultants engaged on a project will also have a contract with the appointing employer. Different forms of contract are used for the appointment of:

- architects – Royal Institute of British Architects
- civil engineers – Association of Consulting Engineers
- project managers – Royal Institution of Chartered Surveyors
- structural engineers – Association of Consulting Engineers
- quantity surveyors – Royal Institution of Chartered Surveyors

The contracts in these circumstances are for the professional services relating to the provision of the construction project. The different forms cover duty of care, fees, copyright, collateral warranties and agreements, professional indemnity insurance, etc. The forms identify materials or substances that should not be used in the construction process. These are: high alumina cement in structural elements, wood wool slabs in permanent formwork to concrete, calcium chloride in admixtures for use in reinforced concrete, asbestos products and naturally occurring aggregates for use in reinforced concrete which do not comply with British Standards 882 or 8110. The fees charged by these consultants will be guided by the fee scales that are published by the various professional bodies. However, in common with the principles of competitive tendering, it is now usual to invite consultants to compete for work, with one of the criteria being the fees charged for the consultancy work.

CHAPTER 5

CONTRACT DOCUMENTS

Introduction

The clauses from the ICE Conditions of contract in this chapter include:

- Contract documents (clauses 5–7)
- Contractor's programme (clause 14)

The contract documents under any civil engineering or building project should include, as a minimum, the following information:

- The work to be performed: This generally requires some form of drawn information. It assists the employer by way of schematic layouts and elevations. Even the non-technical employer is usually able to grasp a basic idea of the engineer's or designer's intentions. Drawings may also be necessary for planning permission. On building projects it is also necessary to obtain building regulations approval. Finally, the drawings will be required by the prospective contractor in order that the designer's intentions can be carried out during estimating, planning and construction. On all but the simplest types of project, therefore, some drawings will be necessary.
- The quality of work required: It is not easy on drawings alone to describe fully the quality and performance of the materials and workmanship expected. It is usual, therefore, for this to be detailed either in a specification or in a bill of quantities. On civil engineering projects it is more usual to include both a bill of quantities and a specification. Under the Joint Contracts Tribunal (JCT) forms of contract that are used almost exclusively for building projects only one of these documents can be a contract document.
- The contractual conditions: In order to avoid any future misunderstandings it is preferable to have a written agreement between the employer and the contractor. For simple projects the conditions appropriate to minor works (see Chapter 19) may be sufficient. On more complex projects one of the more comprehensive forms of contract should be used.
- The cost of the finished work: Capital works projects represent a major investment on behalf of any employer. It is preferable that the costs involved should be as predetermined as possible by an estimate of costs (or tender) from the contractor. On some projects it may only be feasible to

assess the costs involved once the work has been carried out. In these circumstances the method of calculating this cost should be clearly agreed.

- The construction programme: The length of time available for the construction work on site will be important to both the employer and the contractor. The employer will need to have some idea of how long the project will take to complete in order to plan arrangements for the handover of the project. The contractor's costs will, to some extent, be affected by the time available for construction.

Contract documents under the ICE Conditions of Contract

The contract documents on a civil engineering contract typically comprise the following:

- conditions of contract
- specification
- drawings
- bills of quantities
- tender
- written acceptance
- contract agreement (if completed)

These are described in 'Contract' under clause 1 (e) of the ICE Conditions of Contract.

Conditions of contract

The conditions of contract seek to establish the legal framework under which the construction work is to be undertaken. Although the clauses aim to be precise and explicit, and to cover any eventuality, disagreement in their interpretation does occur. In the first instance, an attempt is made to resolve the matter by the various parties concerned. Where this is not possible, it may be necessary to refer the disagreement to alternative dispute resolution or arbitration. The parties to a contract agree to take any dispute initially to arbitration rather than to the courts. This can save time, costs and adverse publicity that may be damaging to both parties. If the matter still cannot be resolved, it is now taken to court to establish a legal opinion. Such opinions, if held, eventually become case law and can be cited should similar disputes arise in the future.

It is always preferable to use one of the standard forms available, rather than to devise one's own personal form. The imposition of conditions of contract which are biased in favour of the employer are not to be recommended. Contractors will tend to overprice the work, even in times of shortage of work, to cover the additional risks involved. Unless there are very good reasons to the contrary, the engineer should attempt to persuade the employer to use one of the standard forms available. The modification of some of the standard clauses, or the addition of special clauses, should only take place in exceptional cases. It is inadvisable to make modifications to the conditions of contract, as the legal results may be different from what was intended. The majority of the standard forms of contract comprise, in one way or another, the following sections.

Agreement

This is that part of the contract which the parties sign. It should be noted that the main contract is between the employer and the contractor. Space is provided for:

the names of the employer
the names of the contractor
the names of other consultants
the date of the signing of the contract
the location and nature of the work
the list of the contract drawings
the amount of the contract sum

In some circumstances it may be necessary or desirable to execute the contract under seal. This is often the case with local authorities and other public bodies. The spaces for the signatures are then left blank and the seals are affixed in the appropriate spaces indicated. After sealing, the contract must be taken to the Department of Customs and Excise, where, upon payment of stamp duty, a stamp will be impressed on the document. Without this the contract will be unenforceable.

ICE Conditions of Contract

The conditions of contract in most of the main forms used for construction contracts (see Chapter 4) have in general a large degree of comparability, but are different in their details. They define the terms under which the work is to be carried out, the relationship between the employer and the contractor. They describe and limit the powers of the engineer. They include, for example, the contractor's obligations (see Chapter 14) in carrying out and completing to the satisfaction of the engineer the work shown on the drawings, in the specification and described in the bills of quantities. They cover matters dealing with the quality of the work, cost, time, nomination of subcontractors, insurances, fluctuations and VAT. Their purpose is to attempt to clarify the rights and responsibilities of the various parties should a dispute arise during the construction of the works.

The conditions of contract used on the majority of civil engineering projects in the United Kingdom are the ICE Conditions of Contract sixth edition. These are issued by the Institution of Civil Engineers, Association of Consulting

Engineers and the Federation of Civil Engineering Contractors. They thus have the agreement of the different parties involved in a contract. Conditions that are biased in favour of one party can be uneconomical. The Conditions of Contract are for use in connection with works of general civil engineering construction.

The ICE Conditions of Contract contain 72 clauses that are grouped under 24 headings or sections. The first edition was published in 1945. Subsequent editions, some with amendments, have resulted in the sixth edition, dated 1991, but reprinted with amendments in November 1995. Other forms or conditions of contract are described in Chapter 4. The ICE Conditions of Contract include a useful index to the Conditions, that is frequently not included with many of the other forms of contract. The conditions also include:

Form of tender
This provides a short description of the works. It also states that the contractor:

- has examined the appropriate documents
- agrees to carry out the works in accordance with these documents
- undertakes to complete the works within the specified time
- will provide security for due performance

The tenders are submitted to the employer or engineer, who will then make a recommendation as to the acceptance of a tender. If the employer decides to proceed with the project, the successful tenderer is invited to have their bill of quantities checked for arithmetical or technical errors. The form of tender may also state that the employer:

- may not accept any tender
- may not accept the lowest tender
- has no responsibility for the costs incurred in their preparation

The form of tender also includes Appendix Part 1 and Part 2. The Appendix to the conditions of contract needs to be completed at the time of the signing of the contract. The completed Appendix includes that part of the contract which is peculiar to the particular project under consideration. It includes information on the start and completion dates, the periods of interim payment, amounts of liquidated damages, etc. The Appendix also includes recommendations of amounts or percentages for some of the information.

Form of agreement
This is the document that both of the parties to the contract will sign. It represents the legal agreement between the parties, i.e. the employer and the contractor.

Form of bond
This provides the details of the bond that is often required by the employer. The performance bond is a document whereby a bank or insurance

company agree to pay a specified sum should the contractor fail to discharge its obligations.

There are also a number of supplementary procedures dealing with:

Conciliation procedure (1994)
Contract price fluctuations
Arbitration

Specification

Specification means the specification referred to in the tender and any modification thereof or addition thereto as may from time to time be furnished or approved in writing by the engineer (clause 1 (g)).

The specification should clearly identify the:

- quality of materials
- standards of workmanship
- samples of materials/finished work that will be required
- tests which are to be applied to the materials and workmanship

The specification describes in detail the work to be executed, the character and quality of materials and workmanship and any special responsibilities of the contractor not covered by the Conditions of Contract. It may also describe:

- the sequencing of the site operations
- a method of construction to be adopted (otherwise this will be determined by the contractor)
- the details of any facilities to be offered to other contractors or subcontractors working on the site

When drafting a specification care should be exercised to avoid any possible conflict with the other contract documents. Where discrepancies do occur between the documents the engineer will be requested to decide on the correct course of action to be followed. This may result in the issue of a variation under clause 52 of the Conditions of Contract.

Many large organisations, such as government departments, adopt standard specifications for their work, that are revised or amended for each project. Such changes are necessary to suit the particular project concerned, changes in practice and the evolution of design and construction methods.

Drawings

This means the drawings referred to in the specification and any modification of such drawings approved in writing by the engineer and such other drawings as may from time to time be furnished or approved in writing by the engineer (clause 1 (g)).

The drawings depicting the civil engineering works should ideally be complete and finalised at tender stage. Unfortunately, this is seldom the case for any construction project and much less so for works of civil engineering construction. This is due in part to the nature of the civil engineering work involved, where projects can involve a large amount of uncertainty at the design stage. Also, employers and engineers rely too heavily upon the clause in the conditions allowing for variations. Employers and engineers will frequently change part of a design to incorporate new ideas and innovation that might not have been available during the pre-tender design stage. Occasionally, it is due to insufficient time being made available for the pre-tender design work. Tenderers should, however, be given sufficient information to enable them to understand what is required in order that they may submit accurate and realistic tenders.

The drawings will include the general arrangement drawings showing the site location, the position of the works on the site, means of access to the site, plans, elevations and sections. Where these drawings are not supplied to the contractors with the other tendering information, the contractors should be informed where and when they can be inspected. The inspection of these and other drawings is highly recommended since it may provide the opportunity for an informal discussion on the project with the designer. Each drawing should include the:

- name and address of the consultant
- drawing number, for reference and recording purposes
- scale – if more than one scale is used they should be of such dissimilar proportions that they are readily distinguishable by sight
- title, which will indicate the scope of the work covered on the drawing

Upon signing the contract, the contractor will be provided with further copies of the contract drawings. These may include copies of the drawings sent to the contractor with the invitation to tender, together with those drawings that have been used in the preparation of the bill of quantities and specification. The list of drawings will be included within the specification. It is usually necessary during the construction phase for the engineer to supply the contractor with additional drawings and details. These may either explain and amplify the contract drawings, or, because of variations, identify and explain the changes from the original design.

The contractor must have an adequate filing system for the drawings. Superseded copies should be clearly marked. They should not, however, be discarded or destroyed until the final cost has been agreed, since they may contain relevant information used for contractual claims. The drawings should be clear and accurate, but because paper expands and contracts, only figured dimensions should be used. Scaling dimensions is therefore very much a last resort.

Bill of quantities

This means a means the priced and completed bill of quantities (clause 1 (h)).

It is desirable that a bill of quantities or measured schedules should be prepared for all types of civil engineering projects, other than those of a very minor nature. The bill comprises a list of items of work to be carried out, providing a brief description and the quantities of the finished work in the project. In conjunction with the other contract documents, it forms the basis on which the tenders are obtained. When priced it allows tenders to be compared. When the contract has been agreed, the rates in the bill of quantities are used to value the work for interim payments and are used to price the actual quantities of work.

A bill of quantities allows each contractor tendering for a project to price on the same information with a minimum of effort. The bill may include firm or approximate quantities, depending upon the completeness of the drawings and other information from which it was prepared. Civil engineering bills of quantities generally, because of the nature of the work involved, assume that the entire project will be remeasured.

Uses

Although the main use of a bill of quantities is to assist the contractor during the process of tendering, it can be used for many other purposes, These include:

- preparation of interim valuations for interim certificates
- valuation of variations that have been either authorised or sanctioned by the engineer
- ordering of materials if care is properly exercised in the checking of quantities and future variations
- preparation of the final account
- future estimating and cost planning
- determination of the quality of materials and workmanship by reference to specification clauses
- use in obtaining domestic subcontract quotations for sections of the measured work
- use as a form of cost data noting the confidentiality of the contractor's prices

Preparation

The bill of quantities should be prepared in accordance with an agreed method of measurement. The Civil Engineering Standard Method of Measurement (CESMM) (third edition), published by the Institution of Civil Engineers, is the preferred document for this purpose. Building projects are generally measured in accordance with the Standard Method of Measurement of Building Works (SMM7), published by the Royal Institution of Chartered

Surveyors. There are also several other methods of measurement available to the construction industry and many countries abroad have developed their own rules for measurement.

The descriptions and quantities are derived from the contract drawings and CESMM, on the basis generally of the quantities of finished work in the completed project.

Contents

The contents of a bill of quantities are briefly as follows:

- Preliminaries and general conditions or preambles: These cover the employer's requirements and the contractor's obligations in carrying out the work. CESMM provides a framework for this section of the bill.
- Measured works: This section of the bill includes the items of work to be undertaken by the main contractor or to be sublet to domestic subcontractors. There are several different forms of presentation that are available for this work. A recognised order for the inclusion of the various items is important, in order to provide quick and easy reference.
- Prime cost and provisional sums: Some parts of the project are not measured in detail but are included in the bill as a lump sum item. These sums of money are intended to cover work not normally carried out by the general contractor (prime cost sum) or work which cannot be entirely foreseen, defined or detailed at the time that the tendering documents are issued (provisional sum). Prime cost sums cover work undertaken by nominated subcontractors.
- Appendices: The final section of the bill may include the tender summary, a list of the main contractors and subcontractors, a basic price list of materials and nominated subcontractors' work for which the main contractor desires to tender.

Contract documents (clauses 5–7)

Documents mutually explanatory (clause 5)

The several documents forming the contract are described as being mutually explanatory. There is no hierarchy of documents, they are considered to be equal in terms of their importance. For example, the drawings are not considered to be more important than the bills of quantities in matters of interpreting the work to be carried out. If ambiguities or discrepancies arise, the engineer will interpret and adjust the contract accordingly. Instructions will be given which are regarded as instructions in accordance with clause 2. Whilst it is not the contractor's responsibility to look for any discrepancies, should any be found they should be brought to the attention of the engineer.

Supply of documents (clause 6)

The successful tenderer will be issued with the following free of charge upon the awarding of the contract:

- four copies of the conditions of contract
- four copies of the specification
- four copies of the bills of quantities (unpriced)
- copies of the drawings as entered in the Appendix to the form of tender and listed in the specification

The contractor must, in turn, supply the engineer with four copies of drawings, specifications and other documents for those parts of the permanent works that are designed by the contractor. Further copies of other drawings, specifications and other documents that are required by the employer are to be supplied by the contractor at the employer's expense.

Copyright

The copyright of drawings, specifications and bills of quantities (other than their pricing) remains with those who prepared them. In the case of permanent works designed by the contractor, whilst they are used for a particular project, they cannot be used on other projects without the permission of the contractor. The engineer has the authority to request adequate information on permanent works designed by the contractor. This can include information necessary for construction purposes or for their proper maintenance after construction.

Further drawings, specifications and instructions (clause 7)

The engineer, under the terms of the contract, is allowed to revise or modify the drawings, specification and other documentation as may be required. There are relatively few projects where some changes to the design or the construction methods to be used do not occur. The engineer must supply copies of this information to the contractor. Where such modifications result in a variation to the works then this shall be deemed to have satisfied clause 51. In good practice, such changes to drawings are normally accompanied by a written engineer's instruction.

Inadequacy of documents

Where, in the opinion of the contractor, the drawings or specifications are not adequate or are insufficient, then the contractor can request the engineer to supply further information in order that the works can be properly constructed. Where there is a delay in the contractor's receiving this information, then this may constitute a reason for granting to the contractor an extension of time under clause 44. Where reasonable, the contractor will also receive additional payments described in clause 52 (4), calculated and paid in accordance with clause 60. The delay may also be due in part to the contractor's own failure to submit information, for example, in the design of permanent works by the contractor. This will be taken into account in respect of determining the amount of delay and any money that is due to the contractor.

Availability of documents on site

One copy of the drawings and specification supplied by the engineer and one copy of all the documents supplied by the contractor are to be kept on the site. These are to be available for use on site by the engineer and other authorised people at all reasonable times.

Permanent works designed by the contractor

Where part of the permanent works is designed by the contractor, the designs must be submitted to the engineer for approval. The designs may include drawings, specifications, calculations and other information. These must satisfy the engineer as to their suitability and adequacy in terms of their design. The information may also include operation and maintenance manuals to enable the employer to operate, maintain, dismantle, reassemble and adjust the permanent works incorporating that design. A certificate of substantial completion for these works will not be issued until the manuals and drawings have been submitted and approved by the engineer.

The engineer is responsible for the integration and co-ordination of the contractor's design with the rest of the works.

Contractor's programme (clause 14)

Whilst the contractor's master programme is not a contract document under the ICE Conditions of Contract, it is useful to consider it in this chapter. It should be noted that under the JCT forms of contract the contractor's programme is a contract document.

Programme

The contractor must within 21 days of the award of the contract provide the engineer with a programme indicating how the works are to be carried out. This must take into account possession of the site and completion of the works which may be in sections. At the same time the contractor must provide for the engineer a general description of the arrangements and methods of construction that are to be used for carrying out the works. If the engineer rejects these proposals a revised programme will be required within a further 21 days of the rejection. In practice, the original contractor's pro-gramme may be revised. It should be remembered that the contractor's price has been determined on the basis of an outline programme prepared at the tender stage. If the engineer takes no action at all within 21 days then the programme is deemed to have been accepted.

Progress

The programme is not used solely to determine that the project can be com-pleted within the allotted time period, but also as a progress measure. Where adequate progress is not being maintained the engineer can request the contractor to redraft or modify the programme to ensure that the work will be completed at the appropriate time determined in the contract. This revised programme must be submitted within 21 days of the request by the engineer.

Revised programme

The engineer can at any time, although not unreasonably, require the con-tractor to submit further information on the methods of construction to be used. This includes the contractor's temporary works and the use of the contractor's equipment. This information may also include the calculation of stresses, strains and deflections that will arise in the permanent works during construction. This is to enable the engineer to decide whether the works can be constructed without detriment to the permanent works. The engineer must then respond within 21 days that the methods suggested are approved or in what respects they are detrimental to the permanent works. Where they are likely to be detrimental the contractor must make changes to meet with the engineer's approval. Once a method of construction has been agreed between the engineer and the contractor, this cannot be changed without the written approval of the engineer. Such consents must not be unreason-ably withheld.

Compliance with the engineer's requirements to change the methods of working may result in delays or costs to the contractor. This may be due to an unreasonable delay on the part of the engineer in approving a method of construction. It may also be due to the application of design criteria by the engineer that could not reasonably have been known at the time of tender. Under these circumstances the contractor may be entitled to an extension of time under clause 44, notice of a possible claim against the employer under clause 52 (4). Profit will be added to the costs involved of temporary and permanent work. Payment is made under clause 60.

The acceptance by the engineer of the contractor's programme in accord-ance with these clauses or the consent by the engineer to the contractor's methods of working does not, of course, relieve the contractor from duties or responsibilities under the contract.

CHAPTER 6

PROCUREMENT STRATEGY

The construction industry

The construction industry embraces the sectors of civil engineering, building and process plant engineering, but the demarcation between these different areas is blurred. It is concerned with the planning, regulation, design, manufacture, installation and maintenance of buildings and other structures. Construction work includes a wide variety of activities depending upon the size and type of projects which are undertaken and the professional and trade skills that are required. Projects can vary from work worth a few hundred pounds to major schemes costing several million pounds and in excess of £1bn. Whilst the principles of execution are similar, the scale, complexity and intricacy can vary enormously. The industry is also responsible for a significant amount of work undertaken overseas on behalf of British consultants and constructors. About 10–15 per cent of the annual turnover of the major contractors is undertaken overseas.

The construction industry has characteristics which separate it from all other industries:

- the physical nature of the product
- the product is normally manufactured on the employer's premises, i.e. the construction site
- projects that are one-off designs and thus no prototype model is available
- the industry has been arranged in such a way that design has normally been separate from construction
- the organisation of the construction process
- the methods used for price determination

The final product is often large and expensive and can represent a client's largest single capital outlay. Buildings and other structures are, for the most part, designed and manufactured to suit the individual needs of each customer, although there is provision for repetitive and speculative work, particularly in the case of housing. The nature of the work also means that an individual project can often represent a large proportion of the turnover of a single contractor in any year.

The construction industry can be measured in several different ways, e.g. turnover, profitability, number of firms, number of employees (see R. C. Harvey and A. Ashworth, *The Construction Industry of Great Britain*, Butterworth-Heinemann, 2nd edn 1997). Typically it is worth about £50bn in Britain. The industry is also diverse. About 60 per cent of its workload is repairs and maintenance and now less than 25 per cent is represented by public sector projects. Civil engineering accounts for approximately 20 per cent of the total output. It has typically accounted for about 6 per cent of Britain's gross domestic product (GDP) and is the fourth largest construction industry within the European Union.

Changes in practice

During the immediate post-war years of the 1950s and 1960s the employers of the construction industry had only a limited choice of procurement methods which they might wish to use for the construction of their projects. The ICE Conditions of Contract were almost uniformly used throughout both the public and private sectors for civil engineering works. The ICE form of contract was similarly used on most major civil engineering projects and had yet to experience competition from elsewhere. The bill of quantities had become the preferred document for use on all medium- to large-sized projects. The somewhat mistrusted cost-plus type arrangements that had been a necessity a few years earlier for the rapid repair of war-damaged buildings and structures were already in decline.

In an age of rapid change the methods in use yesterday are often no longer adequate for tomorrow's activities. Part of this change is due to the development of the computer and the acceptance of its technology in virtually all aspects of commerce and industry. Changes are also occurring in professional attitudes and activities and the once familiar map of skills, knowledge and activities that seemed slowly to evolve is now changing at more regular intervals. These changes are not just symptomatic of the construction industry but are being mirrored elsewhere throughout society. The changes in the construction industry are perhaps being more widely felt since it has much catching up to do compared with its more high-tech cousins. Manufacturing engineering in the United Kingdom has already partially responded to change, largely because its workload had begun to disappear overseas. Had it continued to develop at a relatively slow pace this would have been its eventual fate. UK manufacturing is now tiny compared to its activities in the latter part of the last century and it has exchanged the labour-intensive activities of the past for robotics technology. A comparison of photographs of manufacturing industry at the turn of the century with those of today shows a startling reduction in the levels of manpower. The present-day construction industry has similarities to the car industry of the 1950s. By the end of the current recession, not only will it, too, be much smaller, more of it will be foreign-owned.

The apparent failure of the construction industry to satisfy the perceived needs of its customers, particularly in the way it organises and executes its projects, has been another catalyst for change. In addition, various pressure groups have evolved to champion the causes of the different organisations which may have particular axes to grind. The 1970s oil-price crisis, which had a massive influence on inflation and hence on the borrowing requirements of industry, helped to motivate the construction industry to improve its own efficiency through the way it managed and organised its work. The depression of the 1990s, like previous slumps, encouraged firms of all kinds to attempt to persuade employers to build, by using what were, at that time, innovative approaches to contract procurement.

Changes in methods of procurement

During the period 1970–95, changes in the methods of construction procurement have perhaps been one of the most fundamental changes to have taken place in the construction industry. Comparisons have been made with procurement methods used by other industries and in other countries around the world. Not surprisingly, it is the countries which appear to have been the most successful which have received most attention. The methods adopted by the construction industries in the USA and Japan have been in the forefront of many of our ideas for change.

There has also been a substantial amount of research undertaken for the benefit of employers, consultants and contractors. The many different methods available, together with hybrid arrangements, each have their advantages and disadvantages and no uniform solution has emerged across the construction industry. The choice of procurement method depends upon a combination of the following characteristics and their relative importance:

- type of employer
- size of project
- type of project
- risk allocation
- form of contract to be adopted
- main objectives of the employer
- status of the designer
- relationships with contractors/consultants, such as partnering
- contract documentation required

The following represent some of the procurement trends that are being observed across the entire construction industry:

- Design and construct arrangements are continuing to increase in their popularity
- Traditional contract arrangements often include at least some element of contractor design

- Wider use of forms and conditions of contract designed by the employer and design teams
- Use of minor works forms on projects that are larger than is intended for minor works forms
- Continued use of bills of quantities on larger projects
- Drawings and specifications only continue to be used on small works projects
- Work packages, which require some form of quantification, used with alternative procurement methods
- Decline in the importance of management contracts
- Dramatic reduction in the use of construction management as a method of procurement

The effects of the recent recession in the construction industry may distort the above scenarios since many consultants and contractors have been prepared to undertake work under less than favourable or ideal conditions. The trend also indicates that whilst a variety of procurement methods are being used by different employers, some employers have reverted to the traditional arrangements. Only design and construct is seen as the alternative and competitor to the traditional form of procurement.

General matters

The following are the major issues to be resolved.

Consultants v. contractors

The arguments for engaging a consultant rather than a contractor as the main employer's adviser are inconclusive. The respective advantages and disadvantages may be summarised as follows. Advantages of a contractor-centred approach are said to be:

- better time management
- single-point responsibility
- inherent buildability
- certainty of price
- teamwork
- inclusive design fees

Disadvantages may be:

- problems of contractor proposals matching with employer requirements
- payment clauses
- emphasis may be away from design towards other factors
- employers may still need to retain consultants for payments, inspections, etc.

Competition v. negotiation

There are a variety of ways in which a contractor may seek to secure business. These include:

* speculation
* invitation
* reputation
* rotation arrangement
* recommendation
* selection

Irrespective of the final contractual arrangements made by the employer, the method of choosing the contractor must first be established. The alternatives available for this purpose are either competition or negotiation. Some form of competition in price, time or quality is desirable. All the available evidence suggests that under the normal circumstances of contract procurement the employer is likely to strike a better bargain if an element of competition exists. There are, however, a number of circumstances in which a negotiated approach may be more beneficial to the employer. Some of these include:

* business relationship
* early start on site
* continuation contract
* state of the market
* contractor specialisation
* financial arrangements
* geographical area

The above list is not exhaustive, nor should it be assumed that negotiation would be preferable in all of these examples. Each individual project should be examined on its own merits, and a decision made in the light of the particular circumstances and specific advantages to the employer.

Certain essential features are necessary if the negotiations are to proceed satisfactorily. These include equality of the negotiators in either party, parity of information, agreement as to the basis of negotiation and a decision on how the main items of work will be priced.

Measurement v. reimbursement

There are essentially only two ways of calculating the costs of construction work. Either the contractor adopts some form of measurement and is paid for the work on the basis of quantity multiplied by a rate, or the contractor is reimbursed the actual costs.

A simple drawing and specification contract, for example, relies upon the contractor measuring and pricing the work, even though only a single lump sum may be disclosed to the employer. The measurement contract allows for the payment for risk to the contractor, the cost reimbursement approach does not. Many of the measurement contracts may allow for a small proportion of the work to be paid for under dayworks (a form of cost reimbursement) but it is less common to find cost reimbursement contracts with any measurement aspects. The points to be borne in mind when choosing between measurement or cost reimbursement contracts include the following:

- Contract sum – This is not available with any form of cost reimbursement contract.
- Final price forecast – This is not possible with any of the cost reimbursement methods or with measurement contracts which rely extensively on approximate quantities.
- Incentive for contractor efficiency – Cost reimbursement contracts can encourage wastage, which must then be passed on to the employer.
- Price risk – Measurement contracts allow for this, employers may therefore pay for something which never actually occurs.
- Cost control – The employer has little control over costs where any form of cost reimbursement contract is used.
- Administration – Cost reimbursement contracts require a large amount of clerical work.

Traditional v. alternatives

Until recent times the majority of the major civil engineering projects were constructed using single-stage selective tendering. This method of procurement has many flaws and alternative procedures have been devised in an attempt to address the issues which it raised. The newer methods, or alternative procurement paths, whilst overcoming the failures of the traditional approach, created their own particular problems. In fact, if it were possible to devise a single method which addressed all of the problems, the remaining methods would quickly fall into disuse. In choosing a method of procurement, therefore, the following issues are of importance. They are more fully considered in Chapter 8.

- project size
- costs, inclusive of the design
- time from brief to handover
- accountability
- design, function and aesthetics
- quality assurance
- organisation and responsibility
- project complexity
- risk placing
- market considerations
- financial provisions

Employer's essential requirements

Employers buying a particular service seek to ensure that it fully meets their requirements. The consultant or contractor employed needs to identify with the employer's objectives within the context in which the employer has to operate and should take particular account of any constraints. The following have been identified as some of the important criteria and requirements of employers:

- impartial and independent advice
- trust and fairness in all dealings
- timely information ahead of possible events
- notification of the implications of the interactions of time, cost and quality
- options from which the employer can select the best possible route
- recommendations for action
- good value for any fees charged
- advice based upon a skilled consideration of the project as a whole
- sound ability and general competence
- reliability of advice
- enterprise and innovation

Methods used for price determination

As mentioned above, civil engineering and building contractors are paid for the work which they carry out on the basis of one of two methods: measurement or cost reimbursement.

Measurement

The work is measured in place on the basis of its finished quantities. The contractor is paid for this work on the basis of quantity multiplied by a rate. It is usual, on larger projects, for some form of pre-measurement to take place. This is usually undertaken by the measurement engineer or a representative of the engineer, such as a quantity surveyor. In other circumstances, where bills of quantities are not provided, some form of quantification for pricing purposes will be undertaken by each contractor for tendering purposes. Contractors usually employ quantity surveyors or measurement engineers for this purpose. Where quantities are provided by the engineer they must be sufficiently detailed and comply with a recognised method of measurement. Where they are prepared by the contractor they will be sufficient only to satisfy the particular contractor concerned.

The work may be measured as accurately as the drawings allow prior to the contract being awarded, in which case it is known as a lump-sum contract. Alternatively, the work may be measured or remeasured after it has been carried out. In the latter case it is referred to as a remeasurement (or admeasurement) contract. All contracts envisage some form of remeasurement to take into account variations. Civil engineering projects are usually entirely remeasured. The employer's and the contractor's representatives usually remeasure the work together. Measurement contracts, even where they are on an entirely remeasurement basis, allow for some sort of final cost to be forecast. This offers advantages to the employer for budgeting and cost control purposes. Civil engineering contracts are typically of the remeasurement type, whereas building contracts are more often on a lump-sum basis.

The following are the main types of measurement contracts used in the construction industry.

Drawings and specification

This is the simplest type of measurement contract and is really only suitable for small works or simple projects. In recent years there has been a trend towards using this approach on schemes much larger than was originally intended. Each contractor has to measure the quantities from the drawings and interpret the specification during pricing in order to calculate the tender sum. The method is wasteful of the contractor's estimating resources, and does not easily allow for a fair comparison of the tender sums received by the employer. Interpreting the specification can be a hazardous job even for the more experienced estimator. The contractor has also to accept a greater proportion of the risk, being responsible for the prices as well as for the measurements based upon interpretation of the contract information. Evidence might suggest that contractors tend to overprice this type of work in order to compensate for possible errors or omissions.

Performance specification

This method is a much more vague approach to tendering and the evaluation of the contractors' bids. In this situation the contractor is required to prepare a price based upon the employer's brief and user requirements alone. The contractor is left to determine the method of construction and choose the materials that suit these broad conditions. In practice, the contractor is likely to select materials and methods of construction which satisfy the prescribed performance standards in the least expensive way. Great precision is required in formulating a performance specification if the desired results are to be achieved.

Schedule of rates

With some projects it is not possible to predetermine the nature and full extent of the proposed construction works. In these circumstances, where it is desirable to form some direct link between quantity and price, a schedule of rates may be used. This schedule is similar to a bill of quantities, but without any actual quantities being included. It should be prepared using the rules of a recognised method of measurement (e.g. CESMM 3). Contractors are invited to insert their rates against these items, and they are then compared with other contractors' schedules in selecting a tender. Upon completion of the work it is remeasured in the normal way and the rates are used to calculate the final cost.

This method has the disadvantages of being unable to predict a contract sum, or an indication of the probable final cost of the project. Contractors also find it difficult to price the schedule realistically in the absence of any quantities.

Schedule of prices

An alternative to the schedule of rates is to provide the contractors with a ready-priced schedule, similar to the Property Services Agency's (PSA) Schedule of Rates. The contractors in this case adjust each rate by the addition or deduction of a percentage. In practice, a single percentage adjustment is normally made to all of the rates. This standard adjustment is unsatisfactory since the contractor will view some of the prices in the schedule as high and others as being inadequate to cover costs. It does, however, have the supposed advantage of producing fewer errors in the pricing in the tender documents than the contractors' own price analysis of the work.

Bill of quantities

Even with all the new forms of contract arrangement, the bill of quantities continues to remain the most common form of measurement contract, and the most common contractual arrangement for major construction projects in the United Kingdom (and Commonwealth Countries). The contractors' tenders are able to be judged on the prices alone since they are all using the same qualitative and quantitative data. This type of documentation is recommended for all but the smallest projects.

Bill of approximate quantities

In some circumstances it is not possible to pre-measure the work accurately. In this case a bill of approximate quantities might be prepared and the entire project remeasured upon completion. Although an approximate cost of the project can be obtained, the uncertainty in the design data makes any reliable forecast impossible. Whilst civil engineering projects are typically remeasured, no distinction is made between a bill of quantities and a bill of approximate quantities.

Cost reimbursement

With these types of contractual arrangements the contractor is able to re-coup the actual costs of the materials which have been purchased and the time spent on the work by the operatives, plus an amount to cover the contractor's profit. Dayworks accounts are assessed on much the same basis.

These types of contract are not favoured by many of the industry's employers, since there is an absence of a tender sum and a forecast final account cost. Some also provide little incentive for the contractor to control costs, although different varieties of cost reimbursement build in incentives for the contractor to keep costs as low as possible. They are therefore often used only in special circumstances such as follows:

- emergency work projects
- where time is not available to allow the traditional process to be used
- when the character and scope of the works cannot be readily determined
- where new technology is being used
- where a special relationship exists between the employer and the contractor

Cost reimbursement contracts can take many different forms. The following are three of the more popular types in use. Each of the methods repays the contractor's costs with an addition to cover profits. Prior to embarking on this type of contract it is especially important that all the parties involved are clearly aware of the definition of contractor's costs as used in this context.

Cost plus percentage

The contractor receives the costs of labour, materials, plant, subcontractors and overheads and to this sum is added a percentage to cover profits. This percentage is agreed at the outset of the project. A major disadvantage of this type of cost reimbursement is that the contractor's profits are directly geared to the contractor's expenditure. Therefore, the more the contractor spends on the civil engineering works the greater will be the profitability. Because it is an easy method to operate, this tends to be the selected method when using cost reimbursement.

Cost plus fixed fee

With this method the contractor's profit is predetermined by the agreement of a fee for the work before the commencement of the project. There is therefore some incentive for the contractor to attempt to control the costs. However, because it is difficult to predict the cost with sufficient accuracy beforehand, it can cause disagreement between the contractor's and the employer's own professional advisers. The result is that the fixed fee may need to be revised on completion of the project.

Cost plus variable fee

The use of this method requires a target fee to be set for the project prior to the signing of the contract. The contractor's fee is then composed of two parts, a fixed amount and a variable amount. The total fee charged depends upon the relationship between the target cost and the actual cost. This method provides a supposedly even greater incentive to the contractor to control the construction costs. It has the disadvantage of requiring the target cost to be fixed on the basis of a very 'rough' estimate of the proposed project.

Examples

		£		£
Cost plus percentage	Estimated cost	100,000	Final cost	111,964
	Agreed percentage			
	profit 10%	10,000	10% profit	11,964
	Tender	110,000	Final account	123,928
Cost plus fixed fee	Estimated cost	100,000	Final cost	111,964
	Fixed fee	12,500	Fixed fee	12,500
	Tender sum	112,500	Final account	124,464
Cost plus variable fee	Estimated cost	100,000	Final cost	111,964
	Fixed fee	10,000	Fixed fee	10,000
	Variable fee		Variable fee	
	+/−10%		10% × £11,964	−1,196
		110,000		120,768

The above are examples only. It cannot be assumed that the final accounts would follow these patterns. These are the simple theories of the calculations. In practice the financial adjustments are usually much more complicated.

Procurement management

It is of considerable importance to employers who wish to have civil engineering projects constructed that appropriate advice is provided on the method of procurement to be used. The advice offered must be relevant, reliable and based upon skill and expertise. There is, however, a dearth of objective and unbiased advice available. It is often difficult to elicit the information relevant to a proposed civil engineering project. Construction employers will tend to rely on the advice from their chosen consultant or contractor. The advice provided is usually sound and frequently successful according to the criteria set by the employer. Nevertheless, it may tend to be biased and even in some cases tainted with self-interest. It is perhaps sometimes given on the basis of who makes the first contact with the employer. Methods and procedures have become so complex, with a wide variety of options available, that an improvement in the management approach to the procurement process is now necessary to meet the employer's needs. For instance, the need to match the employer's requirements with the industry's response is very important if customer care and satisfaction are to be achieved. The employer's procurement manager must consider the characteristics of the various methods available and recommend a solution which best suits the employer's needs and aspirations. The manager will need to discuss the level of risk involved for the procurement path recommended for the project under review.

The process of procurement management may be broadly defined to include the following:

- determining the employer's requirements in terms of time, cost and quality
- assessing the viability of the project and providing advice in terms of funding and taxation
- advising on an organisational structure for the project as a whole
- advising on the appointment of consultants and contractors, bearing in mind the criteria set by the employer
- managing the information and co-ordinating the activities of the consultants and contractor through the design and construction phases

The simplistic view is that engineers design and contractors build. Those are their strengths. It is important, however, that someone is especially responsible for the contractual matters that reflect the other aspects of the project.

The effectiveness of a procurement path lies in a combination of the following:

- the correct advice and decision on which procurement path to use
- the correct implementation of the chosen path
- the evaluation during and after its execution

Value for money

Employers are rightly concerned with obtaining value for money. Cheapness is in itself no virtue. It is well worth paying a little more if, as a result, the gain in value exceeds the extra costs. Value for money, in any context, is a combination of subjective and objective viewpoints. Some items can be measured but others can only be left to opinion or, at best, expert judgement. The former can largely be proven, or at least it could if our knowledge were fully comprehensive. The professional's skill is, however, largely in the area of judgement. The procurement method recommended to the employer needs to be that which offers the best value for money. Careful assessment is required to obtain the desired results.

CHAPTER 7

PROCUREMENT SELECTION

Introduction

The execution of a construction project requires both design work and the carrying out of construction operations on site. If these are to be done successfully, resulting in a satisfactorily completed project, then some form of recognised practice must be employed at the outset to deal with their organisation, co-ordination and procedures. Traditionally, an employer who wished to have a project constructed would invariably commission an engineer to prepare drawings of the proposed scheme and, if the scheme was sufficiently large, employ a quantity surveyor to prepare documentation, such as bills of quantities, on which the contractor could prepare a price. These would all be based upon the employer's brief, and the information used as a basis for competitive tendering. This was the system in common use at the turn of the century and still continues to be the most widely used method in practice.

Since the mid-1960s a small revolution has occurred in the way engineers and contractors are employed for the construction of civil engineering projects. To some extent this is the result of initiatives taken by the Ministry of Works in the early 1960s and by the Banwell Committee, which recommended several changes in the way that projects and contracts were organised, one of which was an attempt to try to bring the designers and the constructors closer together. The construction industry continues to examine and evaluate the methods available, and to devise new procedures which address the shortfalls and weaknesses of the current procedures.

There is, however, no panacea for the present difficulties. In fact, it may be argued that change is occurring so fast that a present-day solution may be quite inappropriate for tomorrow. Each of the methods described has its own characteristics, advantages and disadvantages. All have been used in practice at some time; some more than others – largely due to their familiarity and the consequent advantage of ease of application. New methods will evolve in response to current deficiencies and to changes in the culture of the construction industry.

There is also an added issue to consider in attempting to redesign the process and to reinvent methods and procedures from the past. For example, prior to the industrial revolution, designers, engineers and architects would employ individual groups of tradesmen and co-ordinate their activities as a part of the construction process. In more recent years this has been reinvented as management contracting! In order to secure work, engineers, contractors and others employed in the design and construction of civil engineering structures, now employ more aggressive marketing techniques.

Contractor selection

There are essentially two ways of choosing a contractor, either by competition or negotiation. Competition may be restricted to a few selected firms or open to almost any firm who wishes to submit a tender. The options described later are used in conjunction with one of these methods of contractor selection.

Selective competition

This is the traditional and most popular method of awarding construction contracts. The arrangement is shown in Figure 7.1. In essence a number of firms of known reputation are selected by the design team to submit a price. Different codes of procedure have been developed and incorporated within contract documents to improve the process and relationship between employers and contractors during the tender period.

The majority of contract documents state that the employer:

- is not bound to accept any tender
- may choose not to accept the lowest tender
- is not responsible for the cost involved in the preparation of a contractor's tender

Good tendering procedure

Good tendering procedures will take into account the changes in the ways in which projects are procured and ethics or codes of practice. They are usually not mandatory. Employers and contractors want to secure the best possible deal out of a contract. However, enlightened employers and contractors also realise that in order to be effective, particularly over the long term, some principles of fairness should be introduced (see Chapter 9). The following have been identified as good practice:

1. Use of a standard form or set of conditions, rather than one written solely on behalf of one of the parties to the contract. There are clear advantages to all parties in the knowledge that a standard procedure will be followed in inviting and accepting tenders.
2. A limit should be placed on the number of firms invited to tender. Six firms will be able to secure competition in prices. The cost of preparing tenders is considerable and this has to be borne by the industry.

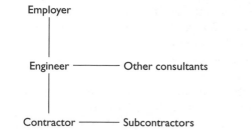

Figure 7.1 Traditional contractual relationship

3. In preparing a short list of tenderers, the following should be considered:
   - the firm's financial standing and record
   - recent experience of constructing over similar contract periods
   - the general experience and reputation of the firm for similar project types
   - adequacy of the firm's management
   - adequacy of capacity
4. Each firm on the short list should be sent a preliminary enquiry to determine its willingness to tender. The enquiry should contain:
   - job title
   - names of employer and consultants
   - location of site and general description of the works
   - approximate cost range
   - principal nominated subcontractors
   - form of contract and any amendments
   - procedure for correction of priced bills
   - contract under seal or under hand
   - anticipated date for possession
   - contract period
   - anticipated date for despatch of tender documents
   - length of tender period
   - length of time tender must remain open for acceptance
   - amount of liquidated damages
   - amount of bond
   - any special conditions
5. Once a contractor has confirmed an intention to tender, that tender should be made. If circumstances arise which make it necessary to withdraw, the engineer should be notified before the tender documents are issued or, at the latest, within 2 days thereafter.
6. A contractor who has expressed a willingness to tender should be informed if not chosen for the final short list.
7. All tenderers must submit their tenders on the same basis.
   - Tender documents should be despatched on the stated day.
   - Alternative offers based on alternative contract periods may be admitted if requested on the date of despatch of the documents.

- Standard conditions of contract should not be amended.
- A time of day should be stated for receipt of tenders and tenders received late should be returned unopened.

8. The tender period will depend on the size and complexity of the job, but should not be less than 4 working weeks, i.e. 20 days.

9. If a tenderer requires any clarification, the engineer must be notified and, in turn, should inform all tenderers of this decision.

10. If a tenderer submits a qualified tender, opportunity should be given to withdraw the qualification without amending the tender figure, otherwise the tender should normally be rejected.

11. Under English law, a tender may be withdrawn at any time before acceptance. Under Scottish law, it cannot be withdrawn unless the words 'unless previously withdrawn' are inserted in the tender after the stated period of time the tender is to remain open for acceptance.

12. After tenders are opened all but the lowest three tenderers should be informed immediately. The lowest tenderer should be asked to submit a priced bill within 4 days. The other two contractors are informed that they might be approached again.

13. After the contract has been signed, each tenderer should be supplied with a list of tender prices.

14. The priced bills must remain strictly confidential.

15. If there are any errors in pricing the Code of Procedure for Selective tendering sets out alternative ways of dealing with the situation.
    - The tenderer should be notified and given the opportunity to confirm or withdraw the offer. If it is withdrawn, the next lowest tenderer is considered. Where the offer is confirmed an endorsement should be added to the priced bills that all rates, except preliminary items, contingencies, prime cost and provisional sums are to be deemed reduced or increased, as appropriate, by the same proportion as the corrected total exceeds or falls short of the original price.
    - The tenderer should be given the opportunity of confirming the offer or correcting the errors. Where the tender is corrected and is no longer the lowest, the next tender should be examined. If it is not corrected then an endorsement is added to the tender.

16. Corrections must be initialled or confirmed in writing and the letter of acceptance must include a reference to this. The lowest tender should be accepted, after correction or confirmation, in accordance with the alternative chosen. Problems sometimes occur because the employer can see that a tender will still be the lowest even after correction. If the first alternative has been agreed upon and notified to all tenderers at the time of invitation to tender, the choice facing the tenderer should clearly be to confirm or withdraw. The employer may require a great deal of persuading to stand by the initial agreement in such circumstances. The answer to the problem is to discuss the use of the alternatives thoroughly with the employer before the tendering process begins. The employer

must be made aware that the agreement to use the Code and one of the alternatives is binding on all parties. It is possible that an employer who stipulated the first alternative and subsequently allowed price correction could be sued by at least the next lowest tenderer for the abortive costs of tendering.

17. The employer is not bound to accept the lowest or any tender and is not responsible for the costs of their preparation. There may be reasons why a decision is taken not to accept the lowest tender. Although the employer is entitled to do so, it will not please the other tenderers. The Code is devised to remove such practices.

18. If the tender under consideration exceeds the estimated cost, negotiations should take place with the tenderer to reduce the price. The quantity surveyor then normally produces reduction or addendum bills. They are priced and signed by both parties as part of the contract bills.

19. The provisions of the Code should be qualified by the supplementary procedures specified in EU directives which provide for a 'restrictive tendering procedure' in respect of public sector construction contracts above a specified value. Guidance on the operation of this procedure is given in DOE Circular 59/73 (England and Wales) and SDD Circular 47/73 (Scotland), both of which are obtainable from HMSO. This method of contractor selection is appropriate for almost any type of construction project where a suitable supply of contractors is available.

Open competition

With this method of contract procurement the details of the proposed project are often advertised in the local and trade publications, or through the local branch of the Construction Employers' Confederation (CEC). Contractors who consider themselves suitable, capable and willing to undertake such a project are requested to write to the engineer for the contract documentation.

This method has the advantage of allowing new contractors or contractors who are unknown to the design team the possibility of submitting a tender for consideration. In theory, any number of firms are able to submit a price. In practice, there is usually a limit on the number of firms who will be supplied with the tender documents. Unsuitable firms are removed from the list if the number of firms tendering becomes too large. The preparation of tenders is both expensive and time-consuming. The use of open tendering may relieve the employer of a moral obligation to accept the lowest price, because firms are not generally vetted before tenders are submitted. Factors other than price must also be considered when assessing these tender bids, such as the capability of the firm which has submitted the lowest tender. There is no obligation on the part of the employer to accept any tender should the employer consider that none of the contractors' offers is suitable. Although open competition is still commonly used in obtaining tenders for minor works its use on larger projects has now largely been curtailed.

Negotiated contract

This method of contractor selection involves the agreement of a tender sum with a single contracting organisation. Once the documents have been prepared the contractor prices them in the usual way. The priced documents are checked for the reasonableness of the contractor's rates and prices and the two parties then meet to negotiate an agreed price for the works. There is an absence of any competition or other restriction, other than the social acceptability of the price. It normally results in a tender sum that is higher than might otherwise have been obtained by using one of the previous procurement methods. Negotiation does, however, have particular applications where:

- a business relationship exists between the employer and the contractor
- only one firm is capable of undertaking the work satisfactorily
- the contractor is already established on site (continuation contract)
- an early start on site is required by the employer
- it is beneficial to bring the contractor in during the design stage to advise on constructional difficulties and how they might best be avoided

Competitive tendering can create financial problems for both the industry and its employers due to excessively keen pricing which in the end benefits no one. Negotiated contracts do result in fewer errors in pricing since all of the rates are carefully examined by both parties. For similar reasons fewer exaggerated claims attempting to recoup losses on cut-throat pricing are likely on negotiated contracts. This type of procurement can also involve the contractor in some participation during the design stage, which can result in savings in on-site time and costs. It should also be possible to achieve greater co-operation during the construction process between the design team and the contractor. The public sector, however, does not generally favour negotiated contracts because of:

- higher tender sums incurred
- public accountability
- suggestion of possible favouritism or collusion

Contractual options

The following contractual options are an attempt to address the employer's objectives in the areas of time, cost and quality of construction. They are not mutually exclusive. For example, it is possible to award a serial contract using in the first instance a design and construct arrangement. Fast tracking may be used in association with a form of management contracting. All of these options will also need to include either a selective or negotiated approach in respect of choosing a contractor.

Early selection

This method is sometimes known as two-stage tendering. Its main aim is to involve the chosen contractor for the project as soon as possible. It therefore seeks to get the firm who knows what to build (the engineer) in touch with the firm who knows how to build it (the constructor), before the design is finalised. The contractor's expertise in construction method can thus be harnessed with that of the engineer to improve constructability criteria in the project. A further advantage is that the contractor may be able to start work on site sooner than if more traditional methods of contract procurement are used. In the first instance, an appropriate contractor must be selected for the project. This is often done through some form of competition and can be achieved by selecting suitable firms to price the major items of work connected with the project. A simplified bill of quantities can be prepared which might include the following items:

- site on-costs on a time-related basis
- major items of measured works
- specialist items, allowing the main contractor the opportunity of pricing the profit and attendance sums

The contractors should also be required to state their overhead and profit percentages. The prices of these items will then form the basis for the subsequent and more detailed price agreement as the project gets under way.

During the first stage it is important to:

- provide a competitive basis for selection
- establish the layout and design
- provide clear pricing documents
- state the respective obligations and rights of the parties
- determine a programme for the second stage

Many of the good practices already outlined for single-stage selective tendering apply equally to two-stage tendering. Acceptance of the first-stage tender is a particularly delicate operation. The employer does not wish to be in the position of having accepted a contract sum at this stage. The terms of the letter of acceptance must be carefully worded to avoid such an eventuality.

After a contractor has been appointed, all unsuccessful tenderers should be notified and, if feasible, a list of first-stage tender offers should be provided. If cost was not the sole reason for acceptance, this fact should be stated.

Design and construct

Design and construct projects aim to overcome the problem of the separation of the designing and constructing processes by providing for these two separate functions within a single organisation (see Figure 7.2). The single

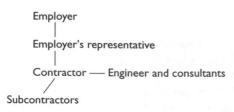

Figure 7.2 Design and construct relationship

firm employed is frequently the contractor. The contractor may employ civil engineers as in-house designers or be responsible for directly employing a firm of consultants. The major difference is that instead of approaching a firm of engineering consultants, the employer briefs the contractor direct, as shown in Figure 7.2.

The employer may choose to retain the services of an independent consultant to assess the contractor's design, to monitor the work on site or to approve payments. The prudent employer will want some form of independent advice. A design evolved by the contractor is more likely to be suited to the needs of the latter's organisation and construction methodology and this should result in savings in both time and costs of construction. (Some argue, however, that the design will be more influenced by the contractor's construction capabilities than the design requirements of the employer.) The completed project should result in lower production costs on site and an overall shorter design and construction period, both of which should provide price savings to the employer. There should also be some supposed savings on the design fees even after taking into account the necessary costs of any independent advice. A further advantage to the employer is in the implied warranty of suitability because the contractor has provided the design as a part of the all-in service. Normally where the ICE Conditions of Contract are used, the contractor has a duty only to use proper skills and care.

A major disadvantage to the employer is the financial disincentive to the employer to make possible changes to the design while the project is under construction. Where an employer considers these to be important to keep abreast of changing technologies or needs, considerably higher costs than is usual may be required either to discourage them in the first place, or to allow for their incorporation within the partially completed project.

Design and construct projects usually result in the employer obtaining a single tender from a selected contractor. Where some form of competition in price is desirable, both the type and quality of the design will need to be taken into account. This can present difficulties in evaluating and comparing the different schemes.

The advantages claimed for a design and construct approach therefore include:

- The contractor is involved from the inception and is thus fully aware of all of the employer's requirements.

- The contractor is able to use specialised knowledge and methods of construction in evolving the design.
- It should be possible to reduce the time from inception to completion due to the telescoping of the various parts of the design and construction processes.
- There can be no claims for delays due to a lack of design information, since the contractor is in overall control.
- There is direct contact between the employer and the contractor.

Package deal

In practice the terms 'package deal' and 'design and construct' are interchangeable. The latter normally refers to a bespoke arrangement for a one-off project. The package deal is, however, strictly a special type of design and construct project where the employer almost chooses a suitable project or building from a catalogue. It is more relevant to building works than to civil engineering works. The employer may be able to view completed projects of similar designs and type that have been completed elsewhere.

This type of contract procurement has been used extensively for the closed systems of industrialised system buildings of timber or concrete. Multi-storey office blocks and flats, low-rise housing, workshop premises, farm buildings, etc., have been constructed on this basis. The employer-owner typically provides the package-deal contractor with a site and supplies the user requirements or brief. An engineer, architect or surveyor may be independently employed to advise on the building type selected, to inspect the works during construction or to deal with the contract administration. The engineer's role is also useful to the employer for those items which are outside the scope of the system superstructure.

It cannot be automatically assumed that this type of procurement will be a more economic solution to the employer's needs, either initially or in the long term. Certainly, the standard system buildings are normally constructed more quickly than the traditional solutions owing to the relative completeness of the design, the availability of standard components, off-site manufacture and the speed of construction on site. However should the employer want to change aspects of the constructional detailing, there is even less scope for variations than with the more usual design and construct approach. Some system buildings constructed in the 1960s are now very costly to maintain.

Design and manage

This method of procurement is really the consultant's counterpart to contractor's design and construct. In this case the design manager, a chartered civil engineer, has full control, not just of the design phase but also of the construction phase. The design and manage firm effectively replaces the main contractor in this role, which in the present day is one largely of management and organisation and the administration and co-ordination of subcontractors. The design manager is responsible for all aspects of construction, including the programming and progressing and the rectification of any defects which may arise. The contract is between the design and manage consultancy and the employer. This provides for the employer a single point of contractual responsibility.

The actual site construction work is generally let through competition in work packages to individual subcontracting firms. This method of procurement therefore offers many of the advantages of traditional tendering coupled with design and construct. The design and manage firm will, of course, need to engage its own site civil engineers or develop those existing staff who have this potential. It will also need to consider continuity with this type of work. The design and manage approach is suitable for all types and sizes of project, but employers undertaking large projects may, due to past experience, prefer a more tried and tested form of procurement using one of the larger contractors. A major disadvantage is in the area of site facilities, which will need to be provided by the design and manage consultant and may have to be hired in a similar way to the subcontractors.

This type of procurement method should be able to give completion times comparable to those offered by the other methods. Since there is the traditional independent control of the subcontractor firms the standard of quality should be at least as good as that provided by the other contracting methods. In terms of cost, design and manage will be no more expensive since the work packages will be sought through competition.

Turnkey method

This form of contracting is still somewhat unusual in the United Kingdom. It has, however, certain notable successes in the Middle and Far East. The true turnkey contract includes everything from inception up to occupation of the finished project. The method receives its title from the 'turning-the-key' concept whereby the employer, on completion of the project, can immediately start using it since it will have been fully equipped (including furnishings) by the turnkey contractor. Some turnkey contracts also require the contractor to find a suitable site for development for this purpose.

An all-embracing agreement is formed with a single administrative company for the entire project procurement process. It is therefore an extension of the traditional design and construct arrangements, and in some cases it may even include a long-term repair and maintenance agreement. On industrial projects the appointed contractor is also likely to be responsible for the design and installation of the equipment required for the employer's manufacturing process. This type of procurement method can therefore be appropriate for use on highly specialised types of civil engineering projects.

The entire project procurement and maintenance needs can thus be handled by a single firm, which accepts sole responsibility for all events. It has been argued, however, that the employer's ability to control costs, quality, performance, aesthetics and constructional details will be very variable and severely restricted by using this procurement method. A contractor who undertakes such an all-embracing project will have a variety of strengths and weaknesses and may well have fixed ideas about the importance of the different aspects of the scheme.

Management contract

Management contracting evolved at the beginning of the 1970s in the United Kingdom with the aim of constructing more complex projects in a shorter period of time and for a lower cost. It may be argued, therefore, that the more complex the project the more suitable management contracting may be. This method is also appropriate to a wide range of medium-sized projects.

The term 'management contract' is used to describe a method of organising the construction team and operating the construction process. The intention is to place the main contractor in a professional capacity to be able to provide the management skills and practical constructing ability for a fee to cover the overheads and profit. The contractor does not, therefore, participate in the profitability of the construction work. The construction work itself is not undertaken by the contractor, nor does the contractor employ any of the labour or plant directly, except with the possibility of setting up the site and those items normally associated with the preliminary works.

Because the management contractor is employed on a fee basis, the appointment can be made early on in the design process. The contractor is therefore able to provide a substantial input into the design, particularly where the practical aspects of constructing the project are concerned. Each trade section required for the project is normally tendered for separately by subcontractors, either on the basis of measurement or a lump sum. This should result in the lowest cost for each of the trades and thus for the construction works as a whole. The work on site needs a considerable amount of planning and co-ordination, more so than in a traditional procurement arrangement. This is the responsibility of the management contractor and an inherent part of the acquired skills. In common with all procurement methods, whilst there are undoubted advantages over its alternatives there are also disadvantages. It is somewhat open-ended, since the price can only be firmed up after the final works package quotation has been received. The later in the contract the work is let, the less time there will be for negotiating price reductions overall without seriously impairing a section of the works.

Management fee

Management fee contracting is a system whereby a contractor agrees to carry out construction works at cost plus a fee paid by the employer to cover the overheads and profit. Some contractors are prepared to enter into an agreement to offer an incentive on the basis of a target cost. This type of procurement is a similar approach to management contracting and cost-plus contracts and has therefore similar advantages and disadvantages.

An alternative approach can also be used where a bill of quantities can be prepared and priced net of the contractor's overheads and profit, or just the profit. These items are then recovered by means of a fee. The system can be as flexible and adaptable as the parties wish. Invariably, as with cost-plus contracting, the fees are generally percentage-related unless some reasonably accurate forecast of cost can be made. Different contracting firms which use management fee contracting have different ways of determining the fee addition. In any case, the total cost is largely unknown until completion is achieved and the records agreed. Overspending is therefore much more difficult to control and any savings in cost tend to be required and made on the later sections of the project.

Construction management contracting

This offers a further alternative procedure to the management contract. The main difference is that the employer chooses to appoint a construction management contractor who is responsible for appointing a design team with the approval of the employer. The employer chooses to instruct the constructor rather than the designer. The construction management contractor is thus in overall project control of both the design and the construction phases. There are similarities between this method and design and construct, though a major difference is that the contractor would invariably appoint outside design consultants rather than choosing to employ an in-house service. The employer may thus form some contractual relationship with these consultants. In reality, construction management contracting is a reversal of the traditional arrangements.

Project management

Although the definitions of these procurement methods can mean different things in different parts of the world or even in different sectors of the construction industry, they are generally those which are understood in the construction industry in the United Kingdom. Project management in this context is a function which is normally undertaken by the employer's consultants rather than by a contractor (see Figure 7.3).

Contractors do undertake project management but in a different context from that of contract procurement. The title of project co-ordinator is similar to and perhaps describes better the role of a project manager. The employer appoints the project manager, who, in turn, appoints the various design consultants and selects the contractor. It is a more appropriate method for the medium- to large-sized project which requires an extensive amount of co-ordination. The function of the project manager is therefore one of organising and co-ordinating the design and construction programmes. Any

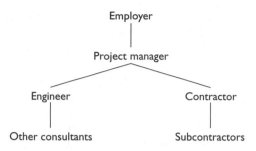

Figure 7.3 Project management relationship

person who is professionally involved in the construction industry can become a project manager; it is the individual rather than the profession which is important.

In general terms the function of a project manager is to provide a balance between function, aesthetics, quality control, economics, and the time available for construction. The project manager's aims are to achieve an efficient, effective and economic deployment of the available resources to meet the employer's requirements. The tasks to be performed include identifying those requirements, interpreting them as necessary and communicating them clearly to the various members of the design team and through them to the constructor. Programming and co-ordinating all of the activities and monitoring the work up to satisfactory completion are also a part of this role. A significant difference between this system and the majority of the others described is that the employer's principal contact with the project is through the project manager rather than through the designer.

Fast tracking

This approach to contracting results in the letting and administration of multiple construction contracts for the same project at the same time. It is appropriate to large construction projects where the employer needs to complete the project in the shortest possible time. The process results in the overlapping of the various design and construction operations of a single project. These various stages may therefore result in the creation of separate contracts or a series of phased starts and completions.

When the design for a whole section of the works, such as foundations, is completed the work is then let to a contractor, who will start this part of the construction work on site while the remainder of the project is still being designed. The contractor for this stage or section of the project will see this work through to completion. While this is being done another work section may be let, continuing and constructing upon what the first contractor has already completed. This staggered letting of the work has the objective of shortening the overall design and construction period from inception through to handover.

This type of procurement arrangement requires a lot of foresight since the later stages of the design must take into account what has already been completed. It will also require considerably more organisation and planning, particularly from the members of the design team. In practice, a project co-ordinator will need to be engaged to ensure an efficient application of this procurement procedure. Although the handover date of the project to the employer should be much earlier than with any of the other methods which might be used, this might be at the expense of the other facets of cost and performance. These aspects may be much inferior to those achieved by the use of the more traditional methods of procurement and arrangement.

Measured term

This type of contract is often used for major maintenance work projects. It is often awarded not just for a single project but to cover a number of different projects of a similar type. The contract will usually apply for a specified period of time, although this may be extended, depending upon the necessity of maintenance standards and requirements and the acceptability of the contractor's performance. The contractor will probably initially be offered the maintenance work for a variety of trades. When completed, the work will be paid for using rates from an agreed priced schedule. This schedule may have been prepared specifically for the project concerned, or it may be a standard document such as the Property Services Agency Schedule of Rates. Where the employer has provided the rates for the work, the contractor is normally given the opportunity of quoting a percentage addition or reduction to these rates. The contractor offering the most advantageous percentage will usually be awarded the contract. An indication of the amount of work involved over a defined period of time would therefore need to be provided for the contractor's better assessment of the quoted prices.

Serial tender

Serial tendering is a development of the system of negotiating further contracts, where a firm has already successfully completed a project for work of a similar type. Initially the contractors tender against each other, possibly on a selective basis, for a single project. There is, however, a contractual mechanism for several additional projects to be awarded automatically using the same contract rates. Some allowances are normally made to allow for inflation, or, perhaps more commonly, increased costs are added to the final accounts. The contractors would therefore know at the initial tender stage for the first project that they could expect to receive a further number of contracts which would help to provide some continuity in their workloads. Conditions would, however, be written into the documents to allow further contracts to be withheld where the contractor's performance was less than satisfactory. Serial contracting should result in lower prices to contractors since they are able to gear themselves up for such work – for example, by purchasing suitable types of plant and equipment that might otherwise be of too speculative a nature. They should therefore be able to operate to greater levels of efficiency. The employer will also achieve some financial gain since some of these lower production costs will find their way into the contractors' tenders. A further advantage claimed for the use of serial tendering is that the same design and construction teams can remain involved for each of the projects which form part of the serial arrangement. It is claimed that this regular and close working association aids and develops an expertise which accelerates the production of the work and eases any anticipated problems. In turn the operatives on site improve their efficiency as they proceed from project to project.

Continuation contract

A continuation contract is an ad hoc arrangement to take advantage of an existing situation. In an expanding economy projects are often insufficient in size to cope with current demands even before they become operational. During their construction an employer may choose to provide additional similar facilities, which may be constructed upon the completion of the original scheme. Such additions, often because of their size and scope, are beyond the definition of extra work or variations. A continuation contract can be awarded as an add-on to the majority of the contractual arrangements. Continuing with the same team (consultants and contractor), where this has already proved to be satisfactory to the employer, is a sensible arrangement under these circumstances. Each of the parties will already be familiar with each other's working methods and will be able to offer some cost savings since they are already established on the site. Both the employer and the contractor may wish to review the contractual conditions and the market factors, which may have changed since the original contract was signed. It is also not unreasonable to expect that the contractor will want to share in the monetary savings that will be available from the already established site organisation from the initial project.

Speculative work (develop and construct)

Because of the market demand, many construction firms have organised a part of their function into a speculative department. These firms or divisions purchase land, obtain planning approvals and design and construct projects for sale, rent or lease. It is also described as develop and construct. It is a more common venture for building projects than in civil engineering. The contractors are often in partnership agreements with financial institutions and developers who have established that the demand for a particular type of development exists in a certain geographical area. Such assumptions are based upon detailed market research of the area undertaken prior to the development getting under way. The developer or speculative contractor may employ their own in-house design staff or may employ consultants to do the work for them. Whilst design and construct is done for a particular employer, speculative construction is carried out on the basis that there is a need for certain types of projects and that these will be purchased at some later date.

Direct labour

Some employers, notably local authorities and large industrial firms, have departments within their organisations which undertake in-house construction work. In some cases they may restrict themselves to repair and maintenance works alone or to minor works schemes. Others are much larger and are capable of carrying out almost any size of construction project. In order to execute construction projects in this way, the employer needs either to have an extensive programme of works or to be responsible for a large amount of repair and maintenance work. In-house services of this type allow the employer, at least in theory, much greater control than using some of the other methods which have been described. In practice, the level of control achieved may be less than satisfactory, due to the accounting procedures which may be employed. Employers who choose to undertake their construction works by this method often do so for either political or financial reasons.

Review

At the inception stage of any construction project both the designers and the contractors should ask themselves about the employer's real objectives. Almost without exception the employer's needs will be an amalgam of:

- functional and aesthetic requirements
- appropriate standards and quality
- completion when required
- appropriate costs
- value for money
- easy and economical maintenance

The contractual procurement arrangements used must aim to satisfy all of these requirements; some in part and others as a whole, trading between them to arrive at the best possible solution. The traditional approach has been to appoint a team of consultants to produce a design and estimate and to select a constructor. The latter would calculate the actual project costs, develop a programme to fit within the contract period, organise the workforce and material deliveries and build to the standards laid down in the contract documentation. In practice, the quality of the workmanship was often poor, costs were higher than expected, and even where projects did not overrun the contract period they took longer to complete than in other countries around the world. Traditionally, little attention or consideration has been given towards future repairs or maintenance aspects.

The obvious deficiencies in the above, supposedly due largely to the separation of the design role from that of the construction effort, have been known for some time. Employers, however, have generally wished to retain the services of an independent designer, believing that such would serve their needs better than the contractors with their own vested interests. They also wished to retain the competition element in order to keep costs down and to help to improve the efficiency of the contractor's organisation. In more recent years they have seen some benefits from constructability, which have been achieved through the early and integrated involvement of the contractor during the design stage. This has also had a spin-off effect in reducing costs and the time that the contractor is on site. However, employers are still loath to commit themselves to a single contractor at the design stage when costs may still be imprecise, and where their bargaining position may be seriously affected before the final deal can be struck.

Increasingly, however, contractors attempted to market their design and construct approach as a possible solution to the employer's dissatisfaction with existing procedures. The benefits of a truly fixed price and a single point of responsibility where the contractor assumed the full responsibility for both the design and the subsequent construction of the project were attractive propositions. Remedies open to the employer in the case of default or delay were now no longer a matter for discussion between the designer and the contractor. If necessary, the employer could always employ professional advisers to oversee the technical and financial aspects of the project.

The entrepreneurs of the construction industry are always seeking out new ways of satisfying the increasing demands of employers. Some are busy examining the methods in use in the USA, and thus seek to import systems which they believe are superior and improve the image of the construction industry. Others feel that the employers' interests would be better served by an overarching project manager or construction manager. Contractors are also going through a major evolutionary change in the way in which they employ and organise their workforce. An increase in subcontracting has occurred as fewer and fewer operatives are directly employed by main contractors. Management contracting recognises and encourages these developments as trends which are desirable.

The construction industry will continue to evolve and adapt its systems and procedures to meet the new demands. New procurement methods will be developed which utilise new technologies and new ways of working. Construction projects are different from the majority of manufactured goods, because they are procured in advance of their manufacture. The majority of

1. Traditional arrangement

2. Early selection

3. Fast tracking

work packages

4. Complete design at tender stage

Key

Design
Construction

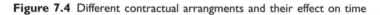

Figure 7.4 Different contractual arrangments and their effect on time

projects are also different from previous ones and often incorporate some new characteristics. Some projects are almost wholly original and all depend upon their peculiar site characteristics. In order to achieve value for money and employer satisfaction, the appropriate procurement procedures must be selected or adapted to suit the individual needs of the employer and the project. Figure 7.4 provides a comparison of typical contract periods using different procurement methods.

CHAPTER 8

PROCUREMENT CONSIDERATIONS

Introduction

The selection of appropriate contract arrangements for any but the simplest type of project is difficult owing to the diverse range of options and professional advice which is available. Much of the advice is in conflict. For example, design and construct contractors are unlikely to recommend the use of an independent designer. Such organisations are also likely to believe that the integration of design and construction is more likely to result in the provision of better civil engineering projects and to improve employer satisfaction. Many of the professions, of course, hold the opposite viewpoint, and believe that their independent approach will produce the best solution, particularly in the long term.

In the past few years there has been a significant shift in the way that construction projects are procured. This is partly in response to the changes which have occurred within the construction industry. Individual experiences, prejudices, vested interests, the desire to improve the system and the familiarity with particular methods are all factors which have influenced capital project procurement recommendations. Those who have had bad experiences with a particular procurement method will be cautious in recommending such an approach again. When it is suggested they will attempt to resist such a course of action, and where the results are as expected, this further reinforces the view that such a method of procurement has only limited use.

In reality too little is known and too little research has been undertaken to evaluate properly the various procurement options. In any case, this is difficult because there are both successful and unsuccessful projects that have used identical means and methods of contract procurement. Personal human factors and external influences also have a major affect upon the possible outcomes. Nevertheless, more reliable information, based on empirical evidence, is desirable if we are to at least attempt to eliminate hearsay, folklore and the apocryphal stories that surround this subject.

The course of action recommended should always be that which best serves the employer's interests, and not that which is perhaps more beneficial or easier to operate for the individual who is offering the advice. This is professionalism, even though it can mean losing a commission. The employer may need to be convinced that a particular method and procedures which are recommended should be adopted as the most appropriate approach. This may at times be against the employer's own better judgement or preferences, particularly where there is some familiarity with other procedures that are being suggested. Some employers are resistant to change and may require persuasion to adopt what might appear to be radical proposals but which nevertheless represent the right advice. It may also be necessary to accept that, because of the mechanisms employed, a better course of action cannot easily be explained until after the event, and that such a recommendation may be rejected at the outset as being speculative.

Overseas employers working in a different cultural environment may find that the traditional contractual procedures used are alien to their needs and expectations. They may require even greater assurances that their project should be constructed using a particular procurement method. Even individuals working within the same team will have different expectations and views on the optimum procurement path.

Procurement advisers should offer advice irrespective of any vested interests or personal gains. Any associated interests which might influence the judgements should be declared in order to avoid possible repercussions at a later stage. Although the professional institutions retain codes of conduct and disciplinary powers governing their members' activities, there has been a blurring of commercial and professional attitudes, the former vying for position with the latter. Outmoded approaches and procedures should be removed when they no longer serve any useful purpose. Sound, reliable and impartial advice is necessary from those who have the skills, knowledge and expertise.

Employer's requirements

A confirmed cynic once described the construction industry as 'the design and erection of buildings and structures that satisfy the designers alone and not the employer'. Furthermore, projects often failed to function and operate correctly, were too expensive and took too long to construct. As in most cynicism the case is vastly overstated, yet there is enough truth in the statement not to dismiss it out of hand.

One of the procurement adviser's main problems is in separating the employer's wants and needs. The employer's main trio of requirements is shown in Figure 8.1. The numerical values result from past research and attempt to weigh these broad objectives in order of their relative importance. They reflect the 'average' or typical employer and offer some guidance on the main issues associated with a construction project. The attempted optimisation of these factors may not necessarily achieve the desired solution. For example, an employer may desire exceptional standards and quality in a project.

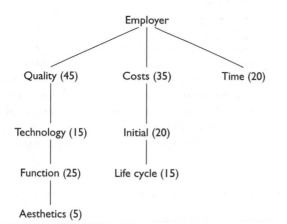

Figure 8.1 Employer's requirements for a typical construction project (weighted values)

To achieve these it may be necessary to incur higher costs or a longer contract period. Alternatively, an employer may require the early use of the project as the main priority and be prepared to sacrifice a bespoke design or even high quality standards in order to achieve this. Employers may also choose to set contrasting objectives which are difficult to achieve using conventional procurement procedures. These may require the adviser to devise ingenious methods of procurement in order to satisfy such objectives.

The employer's objectives are a combination of the following factors. The procurement path will need to prioritise them for each individual project:

- acceptable spatial design to accommodate function and use
- aesthetically pleasing design to the employer (and others)
- final cost of the project should closely resemble the budget estimate
- quality and standards should be in accordance with current expectations
- performance and costs-in-use should be within identified criteria
- available for handover and occupation at the specified time

Factors to consider in procurement selection

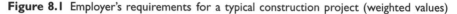

The following need to be considered when choosing the most appropriate procurement path for a proposed project:

- size
- cost
- time
- accountability
- design
- quality assurance

- organisation
- complexity
- risk
- market
- finance

Project size

Small projects are not really suited to the more elaborate forms of contractual arrangements, since such procedures are likely to be too cumbersome and not cost-effective. Smaller-sized schemes therefore rely upon the traditional and established forms of procurement, such as a form of competitive tendering or a version of design and construct. The medium- and large-scale schemes are able to use the whole range of options which are available. On the very large schemes a combination of the different arrangements may be required to suit the project as a whole.

Project size is difficult to quantify. A small civil engineering project might cost up to half a million pounds, a medium one up to ten million and a large one in excess of this.

Cost

Open tendering will generally secure for the employer the lowest possible price from a contractor. Competition helps to reduce costs through efficiency, and lowers the price to the customer. There are limits to how far this theory can be applied in practice. For example, if a large number of firms have to prepare detailed tenders this increases the industry's costs, which must either be absorbed or passed on to the industry's employers through successful tenders. There is no such thing as a free estimate! Negotiated tenders, on the other hand, supposedly add around 5 per cent to the contract price. In the absence of competition, contractors will price the work 'up to what the traffic will bear'.

Projects which require unusually short contract periods incur cost penalties, largely to reflect the demands placed upon the constructor for overtime working and rapid response management. The imposition of conditions of contract which favour the employer or insist upon higher standards of workmanship than are usual also have the effect of increasing the costs of construction. Under these circumstances the employer may end up paying more for stricter conditions which might not be needed or for a standard of workmanship which might not be required.

Employers in the past have been overly concerned with a lowest tender sum, often at the expense of other factors, such as the principles behind the indeterminate life-cycle cost. Cash-flow projections which might have the effect of reducing the timing of expenditures are often ignored except perhaps by the most enlightened employers. A cash-flow analysis, for example, may be able to show that the lowest tender is not always the least expensive solution. The timing of cash flows on a large project may have a significant effect upon the real costs to the employer. With projects that include options on construction method the cash flow analysis becomes even more important. Procurement and contractual arrangements that deal with these factors should be considered more frequently.

The cost of the project is a combination of land, construction, fees and finance, and the employer will need to balance these against the various procurement systems which are available. In terms of cost savings, it may be more appropriate to use a less expensive site elsewhere, if that is possible, rather than reducing other aspects of the project.

Design and construct projects show some form of cost savings in terms of professional fees. The precise amount of these is difficult to calculate since they are absorbed within other charges submitted by the contractor in the bid price. Where the project is of a relatively straightforward design, such as a standard warehouse unit or farm building, it can be more cost-effective to use a contractor who has already completed similar projects, rather than to opt for a separate design service.

Cost reimbursement appears to be a fair and reasonable way of dealing with construction costs. Society is not fair, though, and such an arrangement is often too open-ended for many of the industry's employers. Some employers may choose to use it out of necessity, rating more highly objectives other than cost. The majority of employers, however, will need to know as a minimum the approximate cost of the scheme before they begin to build, and there is nothing new in this (St. Luke's Gospel 14: 28).

Where a firm price is required before the contract is signed one of several procurement arrangements can be used. A firm price arrangement relies to a large extent on a relatively complete design being available. Where more price flexibility in the design is required, one of the more advanced forms of procurement might be used, while at the same time achieving a measure of cost control.

Provisions exist under most forms of contractual arrangement for a fixed or fluctuating price agreement. The choice is influenced by the length of the contract period and the current and forecast rates of inflation. Where the inflation rate is small in percentage terms and falling, then a fixed-price arrangement is preferable. When the rate of inflation is small and stable it is common to find projects of up to 36 months' duration on this basis. When the rate of inflation is high – and particularly when it is rising – contractors will be reluctant to submit fixed-price tenders for more than about one year's duration. Although a fixed price is attractive to the employer, it cannot be assumed that this will necessarily be less expensive than a fluctuating type arrangement. Some contract conditions limit fluctuation reclaims to increases caused directly by changes in government legislation.

It is difficult to make cost-procurement comparisons, even where similar projects are being constructed under different contractual arrangements. The following are the cost factors to consider:

- price competition/negotiation
- fixed-price arrangements
- price certainty
- price forecasting

- contract sum
- bulk-purchase agreements
- payments and cash flows
- life-cycle costs
- cost penalties
- variations
- final cost

Time

The majority of employers, once they have made the decision to build, want the project to be completed as quickly as possible. The design and construction phases in the United Kingdom are lengthy and protracted. Some of the apparent delays are linked to the protracted planning processes rather than the design or the construction phases. It is difficult to make comparisons on a global basis since there are a wide variety of influences to be considered, such as methods of construction, safety, organisation of labour, quality assurance, etc. Construction techniques also vary and these, in turn, produce different qualities, costs and time periods. The time available will also influence the type of construction techniques which might be used. A need for rapid completion may force the employer to consider using an 'off-the-peg' type project that can be constructed quickly.

Several different methods of procurement have been devised with the objective of securing the early completion of the project. Such approaches have implications for the other factors under consideration such as design, quality or cost. There is an optimum time solution, depending upon the importance which is attached to these other considerations. For example, in terms of cost, shorter or longer periods of time on site tend to increase construction costs. The former is due in part to overtime costs, the latter to extended site on-costs. Some employers are prepared to pay extra costs in order to achieve earlier occupation. Different techniques of design and construction may also need to be used to achieve early completion.

Research has indicated that construction work should not commence on site until the project has been fully designed. In this case the project is able to be constructed with minimal involvement thereafter by the designer. Variations will still be allowed, but are not encouraged. Such an approach helps to eliminate a large amount of construction uncertainty which is common on many supposedly designed projects today. This approach allows the contractor to plan the organisation and management of the project better and to spend less time awaiting drawings and details. Overall, the design and construction period is shortened and earlier completion is achieved. The contractor is on site for a shorter period of time, which results in the saving of construction costs. The difficulty of adopting this approach is that most employers, once they are committed to a project, want the work on site to commence as soon as possible.

Many of the newer methods of contract procurement have been devised specifically to find a quicker route through the design and construction processes. In some ways they have been assisted by changes in the techniques used for the construction and assembly of the construction products and materials. Early selection was developed to allow the construction work to start on site while some of the scheme was still at the design stage. It also allowed the contractor an early involvement with the project. Some form of cost forecasting and control can be used but these will be much less precise than with some of the more conventional methods of procurement. Critical path analysis can be used to find the quickest way through the construction programme, and the American fast-track system was imported for the sole purpose of securing an early completion of the project. Inaccurate design information, coupled with a need for speed in completion of the works, has often resulted in abortive parts of the project, poor quality control and higher costs to both the contractor and the employer. The later forms of management contracting offer some solution to the time delay problem. Project management contracts, where an independent organisation controls both the design and the construction teams, have provided some good examples in terms of project co-ordination.

The following are the time factors to consider:

- completion dates
- delays and extensions of time
- phased completions
- early commencement
- optimum time
- complete information
- fast tracks
- co-ordination

Accountability

According to the dictionary definition accountability is 'the responsibility for giving reasons why a particular course of action has been taken'. In essence it is not simply having to do the right things, but having to explain why a particular choice was made in preference to others that were available. It is of greater significance when dealing with public employers, where it is necessary to justify why a particular course of action was taken. It has also become increasingly important with all types of employers where an emphasis is placed on achieving value for money on capital works projects. The documentation used for construction works is often complex and the technical and financial implications are considerable.

Employers need the assurance that they have obtained the best possible procurement method to suit their list of objectives. The possible trade-offs between competing proposals will need to be evaluated. It is difficult to satisfy the accountability criteria in respect of price where tenders are sought in the absence of any form of competition. There is also the difficulty of justifying subjective judgements where these appear to be in conflict with common practice. The process of contractor selection will never be a solely mechanistic process. The elimination of procedural loopholes should be such as to provide the employer with as much peace of mind as is possible.

Accountability is interlinked with finance and an emphasis on paying the lowest price for the completed project. It may be easy to demonstrate to some employers that to pay more for a perceived higher quality or earlier completion is worthwhile. Other employers may need more convincing and some will feel doubtful about non-monetary gains.

The procedure for the selection, award and administration of contracts must be as precise as possible. Auditing plays a useful role in the tightening up of the procedures used, with ad hoc arrangements that breach these procedures being discouraged. Systems that require huge amounts of documentation and the subsequent checking and cross-checking of invoices, time sheets, etc., are not favoured because of loopholes that can occur. Open-ended arrangements which are unable to provide a realistic estimate of cost are fraught with problems in terms of accountability. They present difficulties in demonstrating value for money at the tender stage, since any forecast of cost will be too imprecise. The following are the accountability factors to consider:

- contractor selection
- ad hoc arrangements
- contractual procedures
- loopholes
- simplicity
- value for money

Design

There is a good argument that the best design will be obtained from someone who is a professional designer. The design will then not be limited by the capabilities of the constructor or restricted to those designs which might be the most profitable to such a firm. However, the design and construct contractor is more likely to be able to achieve a solution which takes into account constructability and produces a design which is sound in terms of its construction. There are many examples where the employer, having been provided with an unsatisfactory construction, has to wait impatiently whilst the consultant and constructor argue about the liability. There are also examples where design and build projects, some using industrialised components, have had to be demolished after only a few years of life, because of their poor design concept, impossible and costly maintenance problems and the unacceptable user environment which they have helped to create.

Contractors have, on the whole, been better at marketing their services and these have reaped benefits in the growing increase in design and build schemes. Designers' response to this upsurge in activity has to some extent been limited by the restrictions imposed on advertising by their profession and by a failure to react to the changes that have taken place. Only in recent years has the design and manage approach been introduced.

The traditional methods of contract procurement fail in many aspects of construction design – not least because of the absence of any constructor input. However, they still represent the most common method of procurement and more publicised failures might thus have been expected. Some of the forms of management contracting, which still largely retain the independence of the designer, did gain some popularity amongst employers, but are now in decline. Where the employer has been encouraged to form a contractual relationship with a single organisation, then this has on the whole been beneficial.

Designs which evolve and develop only marginally ahead of the construction works on site must be of questionable worth in terms of their design solution. In some circumstances this approach is necessary when working within the constraints of either previously completed schemes or the confines of an existing structure. The following are the design factors to consider:

- aesthetics
- function
- maintenance
- constructability
- contractor involvement
- standard design
- design before construct
- design prototypes

Quality assurance

Open tendering can result in a lower standard of workmanship than might have been achieved by using a contractor who submitted a higher price. The statement 'you only get what you pay for' is true in terms of construction quality. Where a contractor has had to submit an uneconomical price standards and quality may suffer if good site supervision is not provided. Consistent and good quality control procedures are often lacking in the construction industry.

The quality of projects depends upon a whole range of inputs including the soundness of the design, a correct choice of specification, efficient working details, adequate supervision and the ability of the contractor. The skills of the operatives are also important, perhaps even more so today than a decade ago, as designs are tending to become more complex in their detailing and higher levels of craftsmanship are expected. The choice of a contractor who has a good reputation for the type and quality of work envisaged is important in achieving this objective. The use of labour-only operatives and the general subcontracting phenomena have sometimes resulted in a deterioration in quality performance, due to poor co-ordination and supervision by site management.

The quality of design and construct schemes depends largely upon the reputation of the contractor selected, particularly where the employer chooses not to involve any professional advisers. The quality of the materials and workmanship will be regulated entirely by the contractor. Speculative construction schemes where the quality assurance is determined solely by the contractor are not renowned for their high-quality work. Fast-track procurement methods which can involve a number of contractors on the same project can, without adequate supervision, result in widely varying standards of workmanship. There is also the difficulty of co-ordinating different contractors on the site, as occurs with work packages on management contracts. The use of selective tendering, properly managed, continues to offer a good solution in terms of quality assurance.

Standards and quality cannot be judged at the construction's completion alone, but need to be considered in the longer term. A virtually complete design prior to the commencement of work on site is likely to be beneficial in improving the qualitative aspects of the project rather than the more ad hoc design approach to problems as they occur on site.

The turnkey method where the designer-contractor has a contractual responsibility for the long-term repair and maintenance of the project offers advantages, particularly in terms of quality assurance. Under this method of procurement there is an incentive for the designer-contractor to reduce the likelihood of future defects arising by a more careful design and effective site management during construction. This may result in less innovation but this is preferable to inconvenience and costly failures in the future. Progress in construction is necessary, but not at the expense of prototype designs which result in poor quality assurance.

Although serial and continuation contracts, using the same design and detailing, should improve the quality aspects, in practice this has not always been achieved. Sometimes, poor design, detailing and construction methods have, unfortunately, been repeated. The following are quality-assurance factors to consider:

- quality control
- independent inspection
- team working
- co-ordination
- subcontracting
- constructability
- future maintenance

- design and detailing
- reputation of craftsmen

Organisation

Allowing the contractor total control of the construction project, as in design and construct or management contracting, removes a layer of organisation and eliminates dual responsibility. This should result in fewer things being overlooked or forgotten, work left undone or subcontractors being unable to complete their work on time owing to a lack of information. The additional tier of organisation has the disadvantage of the parties blaming each other when disputes arise. However, the traditional methods of contract procurement have to a large extent set the lines of demarcation between designer and constructor quite clearly. In practice the designer probably relies too heavily on the constructor. Elaborate conditions of contract, to cope with the organisation of construction work, have had to be drawn up to anticipate most of the eventualities that might arise.

The employment of a single firm, such as a design and construct contractor, allows for quick response management, the ability to deal with problems as they occur, and more freedom in the execution of the works. The extent of such freedom will vary with the conditions of contract being used. Where a separate designer is used the response time is often much longer and this can result in delays to the contract. Where complex or difficult contract arrangements are employed these can have the effect of removing the initiative from the contractor.

The more parties involved with a construction project, the more complex will be the organisation and the contractual arrangements. The employment of a group design practice should therefore result in fewer organisational difficulties than where separate consultants are employed. Management contracting is based upon awarding individual work packages to a range of specialist and general subcontractors. This can create problems of organisation and co-ordination. There is much less control over such firms than when directly employed operatives of the main contractor are used. General contracting is, however, now unusual and about 90 per cent of construction work is subcontracted regardless of the method of procurement being used. These individual firms need to be programmed for precise periods of time, and a delay in allowing them to proceed with their work or a failure by them to complete on time can have a knock-on effect for the whole project. This presents even greater difficulties with tightly scheduled construction programmes. The following are the organisational factors to consider:

- complexity of arrangements
- single responsibility
- levels of responsibility
- number of individual firms involved
- lines of management

Complexity

Projects which are complex in design or construction require more precise and comprehensive contractual arrangements. Complexity may be due to an innovative design, the utilisation of new constructional methods, the phasing of the site operations or the necessity for highly specialised work. It can also be the result of employing several contractors on the same site at one time in order to achieve rapid progress or the complicated refurbishment of an existing construction while still in use by its occupants. It is often necessary in circumstances of these types to devise new contractual arrangements and to apply different types of procedures to the varying parts of the construction work.

Where work can be reasonably well defined and forecast, traditional estimating processes can be used and the work paid for on the usual basis. Where the work is indeterminate, of an experimental type or requires a solution from the contractor, a lump-sum or cost reimbursement approach with contractor design may need to be employed. In the latter case the contractor is given the opportunity to offer an acceptable solution to the problem as a part of the contract. Where the project is very complex, the employer is likely to choose a separate designer with the skills required to produce the right solution. It is, however, important to involve the constructor in the project as soon as possible, particularly where this might influence the sequencing of site operations. A form of two-stage tendering might therefore be appropriate to cope with this eventuality. The following are the organisational points to consider:

- nature of complexity
- capabilities of parties
- main objectives of employer

Risk

Risk is inherent in the design and construction of a civil engineering project. One of the employer's intentions is to transfer as much as possible of the risk to either the consultants or the contractor. Risk may be defined as possible loss resulting from the difference between what was anticipated and what actually occurred. Risk is not entirely monetary. An unsatisfactory design, even if completed successfully, can result in a weakening of the designer's reputation with a consequent loss of future commissions. Risk can be reduced but it is difficult to eliminate it entirely. For example, the risk associated with a very specialised form of construction can be reduced by selecting a contractor with the appropriate experience.

The transfer of risk from the employer to others involved with the project may appear to satisfy the accountability criteria. It may be argued as an appropriate course to follow for the employer, but it may not be a fair and reasonable approach. Nor is it necessarily the best route for the employer to follow, since the risk needs to be evaluated. All contractors' tenders contain a premium to cover contractual risk. Where the risk does not materialise, this becomes a part of the contractor's profit. The employer may thus be better advised to assess and accept some of the risk involved, thereby reducing the contractor's tender sum and also costs accordingly. This is a more common way of dealing with risk on construction projects.

The lump-sum contract with a single price, which is not subject to any variation, is at one extreme and at the other is cost reimbursement, where risk and financial predictability are uncertain. In the former the employer is paying for eventualities which might not occur. In the latter the employer is accepting the risk, but only pays for events that happen. A balance has to be struck. Risk should always be placed with the party to the contract who is in the best position to control it. Where this is not possible then it should at least be shared, although it may be difficult to convince the employer that this course of action is the most financially appropriate. Some projects involve a large amount of risk in their execution. In some cases the risk may be so high that it is impossible to get a contractor even to consider tendering under conventional arrangements. Some form of risk sharing may then become essential in order for the project to proceed. The following are risk factors to consider:

- risk evaluation
- risk sharing
- risk transfer
- risk control

Market

The selection of a method of procurement will be influenced by the state of a country's economy and the industry's workloads. An appropriate recommendation for today may have different implications for some time in the future when economic performance is different.

When there is ample work available, contractors are able to choose those schemes which are the most financially beneficial. Under these circumstances, employers may be unable to insist upon onerous contractual arrangements and conditions of contract. Where the risk involved in executing the work is high it will be even more difficult to persuade contractors to tender for the work. Employers may need to be advised to delay their construction projects at such times, and to wait until the economy becomes more favourable. Many employers will, however, be unable or unwilling to delay their projects.

When construction prices are low, then a form of cost reimbursement or management fee approach can be expensive. On these occasions, contractors are sometimes prepared to undertake work at cost, and take a risk that nothing of any financial significance will go wrong. Conversely, in times of full order books the opposite is true, and paying contractors their actual costs plus an agreed amount for profit may be a better proposition.

When work is plentiful, contractors sometimes have difficulty in recruiting a competent workforce of skilled operatives and professional staff and this, coupled with similar restrictions in the availability of good supervision, can result in a deterioration in quality standards. When the amount of construction work available is restricted, then the greater availability of skilled people coupled with more intensive inspection, is likely to enhance the overall quality of the project. The following are the market factors to consider:

- availability of work
- availability of contractors
- economy effects
- procurement advice

Finance

The usual way of paying the contractor for the construction work is through monthly or stage payments. These payments help the contractor to offset the financial borrowing that is required to pay wages, salaries, goods, components and materials. There are two major alternatives to this common procedure. The first is a delayed payment system similar to that used on speculative developments. The employer effectively pays for the work by way of a single payment upon completion of the project. The employer has to accept the design as it is built, but acquires immediate occupation of the project. The financial borrowing requirements of the contractor are higher, but the employer makes savings by paying for the work only at the end of the project.

The second alternative is for the employer to fund the work in advance and thereby reduce the contractor's interest charges that are otherwise included in the tender. In these situations there are possible problems of which the employer needs to be aware. The industry is notorious for the number of insolvencies, and the employer would thus want to ensure the financial soundness of the appointed contractor. Contractors also tend to be less interested in a project once they have received payment for work. The employer can devise remedies to deal with these factors. With the former, a performance bond can be adopted and with the latter, liquidated damages can be applied. The following are financial factors to consider:

- payment systems
- financial soundness of parties
- financial remedies
- contract funding

Conclusions

The choice of a particular method of contract procurement for a construction project involves identifying the employer's objectives, balancing these with the procurement methods which are available and taking into account

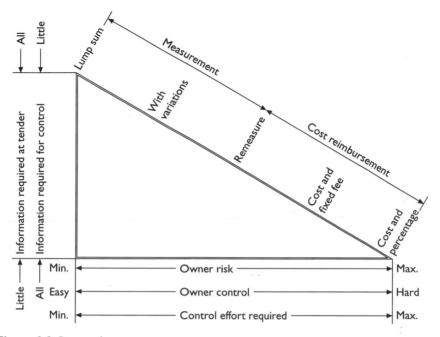

Figure 8.2 Range of contract types
Source: adapted from R.A. Burgess (ed.), *Construction Projects, their Financial Policy and Control,* Longman, 1980

the considerations outlined above. Figure 8.2 provides a comparison of some of the payment methods. It compares factors such as the information requirements with the control and risk considerations. For instance, if an employer's main priority is a lump sum price from a contractor, then full information must be provided at the tender stage for an accurate price to be prepared. Risk and control effort on the part of the employer throughout the duration of the contract will then be minimal. On large and complex projects it is difficult to provide this detailed information. The quality and reliability of the design information will determine how precisely the construction's costs can be forecast and controlled. The poorer and more imprecise it is, the greater will be the risk to the employer. The risk, of course, may never materialise and hence there will be no loss to the employer.

The three broad areas of concern to the employer, and the employer's requirements of any construction project have already been identified in Figure 8.1. The balance of these will vary but their analysis will help to influence and select the most appropriate procurement method.

Table 8.1 offers a checklist of questions to help to determine an appropriate contract strategy. It provides examples of some of the more usual contractual arrangements which are available. It does, however, need to be emphasised that the solutions recommended are based largely upon judgement rather than objective analysis. Under differing circumstances or where other factors

Table 8.1 Identifying the client's priorities

		Traditional selective tendering	Early selection	Design and build	Construction management	Management fee	Design and manage
TIMING	– Is early completion important to the success of the project?						
	Yes:		✓	✓	✓	✓	✓
	Average:		✓	✓	✓	✓	✓
	No:	✓					
VARIATIONS	– Are variations to the contract important?						
	Yes:	✓	✓		✓	✓	✓
	No:			✓			
COMPLEXITY	– Is the project technically complex or highly serviced?						
	Yes:	✓	✓		✓	✓	✓
	Average:		✓	✓	✓	✓	✓
QUALITY	– What level of quality is required?						
	High:	✓	✓	✓	✓	✓	✓
	Average:	✓	✓	✓	✓	✓	
	Basic:						
PRICE CERTAINTY	– Is a firm price necessary before the contracts are signed?						
	Yes:	✓	✓	✓	✓	✓	✓
	No:						
RESPONSIBILITY	– Do you wish to deal only with one firm?						
	Yes:		✓	✓		✓	✓
	No:	✓			✓		
PROFESSIONAL	– Do you require direct professional consultant involvement?						
	Yes:	✓	✓		✓	✓	✓
	No:			✓			
RISK AVOIDANCE	– Do you want someone to take the risk from you?						
	Yes:	✓	✓	✓			
	Shared:				✓	✓	✓
	No:						

Source: adapted from *Thinking about Building*, EDC Report.

need to be taken into account, the solutions must be adjusted accordingly. The questions themselves are not weighted and users will need to do this in order of importance. Some employers may wish to emphasise only a single aspect such as quality and choose a method and contractor which are capable of securing this, but the majority of employers are interested in an amalgam view, trading off the various factors against each other. It is inappropriate to use the chart in an incremental fashion by adding the various answers together.

Choosing the correct procurement method is a difficult task owing to the wide variety of options which are available. Some of the changes in methods of procurement are the result of a move away from the craft base to the introduction of off-site manufacture, the use of industrialised components and the wider application of mechanical plant and equipment. The improved knowledge of production techniques, coupled with the way in which the workforce is organised, has enabled the contractor to be able to analyse the resources involved and move towards their greater optimisation. Contractors also have a much greater influence upon the design of the project and the recognition of constructability has influenced both the design and the way the work is carried out on site and hence the quality of the finished work. The time available for construction and the subsequent costs involved have also been affected by these changes.

CHAPTER 9

PROCUREMENT ISSUES

Introduction

Traditionally, employers who wished to have projects constructed would invariably commission a designer, normally a civil engineer for civil engineering projects and an architect for building projects. These would prepare the working drawings for the proposed project. On civil engineering works a specification would also be provided and, where the project was of a sufficient size, bills of quantities would be prepared. The contractors invited to tender for the works would then prepare their prices using comparable information. This was the common procedure during the early part of the century and still remains the most popular method for the procurement of construction projects. Even up to thirty years ago there was a limited choice of methods of contract procurement. However, since the early 1960s there have been several catalysts for changes in the way that projects are procured. These are shown below.

- Government intervention through committees, such as the Banwell Reports of the 1960s and more recently through the Department of the Environment and the Latham Report (1994)
- Pressure groups formed to encourage change for their members, most notably the British Property Federation
- Large employers of the industry who have been able to develop their own contractual procedures, such as the Department of Transport
- International comparisons, particularly with the USA and Japan and influence of the Single European Market in 1992
- The apparent failure of the construction industry to satisfy the perceived needs of its customers, particularly in the way in which it organises and executes its projects
- Influence of educational developments and research
- Trends towards greater efficiency
- Rapid changes in information technology, both in respect of office practice and manufacturing processes
- Attitudes amongst the professions
- Employers' desire for single point responsibility

There is no panacea and procedures will continue to evolve in order to meet new circumstances and situations. Procurement methods of a hybrid nature are being developed in an attempt to utilise the best practice from the various competing alternatives. All the different methods have been used at some time in the industry; some more than others, due largely to:

- user familiarity
- ease of application
- recognition
- reliability

New procurement systems will continue to be developed to meet new requirements and demands from employers, contractors and the professions.

Published reports

Several different reports, many of which have been government-sponsored, have been issued over the past fifty years. Their overall aim has been to improve the way that the industry is organised and the way that construction work is procured. These various reports are listed below.

> *The Placing and Management of Building Contracts*, 1944 (The Simon Report)
> *A Code of Procedure for Selective Tendering*, 1959
> *Survey of Problems before the Construction Industry*, 1962 (The Emmerson Report)
> *The Placing and Management of Contracts for Building and Civil Engineering Works*, 1967 (The Banwell Report)
> *Action on the Banwell Report*, 1967
> *Construction Contract Arrangements in European Union (EU) Countries*, 1983
> *Building Towards 2001*, 1991
> *Trust and Money*, 1993
> *Constructing the Team*, 1994 (The Latham Report)

Constructing the Team (The Latham Report)

This report provides a review of the construction industry. It was jointly commissioned by the government and the industry. It also included contributions from various employer bodies on their expectations when undertaking major capital works programmes.

The Joint Review of Procurement and Contractual Arrangements in the United Kingdom Construction Industry was announced in the House of Commons in 1993. The interim report, entitled, *Trust and Money* was published later that year. The funding parties to the Review included the Department of the Environment (DOE), the Construction Industry Council (CIC), the

Construction Industry Employers Council (CIEC), the National Specialist Contractors Council (NSCC) and the Specialist Engineering Contractors Group (SECG). Employers have continued to be closely involved in the Review and were represented by the British Property Federation (BPF) and the Chartered Institute of Purchasing and Supply (CIPS).

The interim report sought to:

- describe the background to the Review and its parameters
- define the concerns of the differing parties to the construction process, some of which were mutually exclusive
- pose questions about how performance could be improved and genuine grievances or problems assessed
- reiterate and expand upon what has long been accepted as good practice in the industry, but is often honoured more in the breach than in the observance

It has already been noted that several government-sponsored studies of the organisation and management of the construction industry have been undertaken this century. Whilst the industry has responded in some measure to change, this is attributable to pragmatic courses of action and commercial pressures rather than to the advice offered by government or industry committees.

It is accepted that the construction industry is fragmented and comprises many different parties, organisations and professions with a variety of vested interests and traditions that, in some cases, represent power and authority. Such bodies are clearly loathe to relinquish these positions freely. Whilst there is agreement that a common set of standard conditions of contract could be desirable, there is also resistance to change. This resistance comes from those responsible for the existing forms and conditions of contract. In many cases they believe that the procedures they have prepared offer advantages over those of their possible competitors. There is also the fact that the different forms and conditions of contract represent a commercial consideration. Users, often due to familiarity with a particular set of conditions, are reluctant to change to new procedures, even where they recognise flaws in particular practices. Some civil engineers continue to implement successfully the now outdated fifth edition of the ICE Conditions of Contract in preference to using the sixth edition, first introduced in 1991.

Some recommendations of the various government reports have been implemented voluntarily, others have been introduced through legislation. Difficulties continue to persist, even though both the structure of the industry and its employers have changed dramatically.

Employers are at the core of the process and their needs must be met by industry. Employers are not represented by a single organisation. They frequently have different aims and aspirations. Government, a major employer, in the past used to act as a monolithic employer. Many of its good practices were introduced and accepted by other employers. However, the privatisation of many government departments and activities has changed this perception, resulting in the fragmentation of this important employer base. Even existing government departments now operate different procurement strategies and practices. This became more pronounced, for example, after the demise of the Property Services Agency (PSA) and the privatisation of the different utility companies.

Fair construction contracts

The Construction Sponsorship Directorate of the Department of the Environment published a consultation paper under the above title in May 1995. The review recognised that present arrangements used in the industry militate against co-operation and teamwork and therefore against the employer's requirements. They also help to perpetuate the generally poor image of the construction industry.

This consultation paper identifies four areas which it describes as essential terms in any fair construction contract:

- Dispute resolution: emphasising the need to adopt ADR procedures (see Chapter 3)
- Right to set off: This principle is widely applied beyond the remit of the construction industry. The principle is as follows; where A claims money from B, B is entitled to deduct any money that A owes before paying the balance. The Latham Report recommends that the right to set off should be restrained as follows:
  - a requirement to give advance notification (with reasons) of the intention to apply set-off.
  - set-off subject to adjudication
  - set-off only to be allowable in respect of work covered by the contract
- Prompt payment: with added interest in the case of default
- Protection against insolvency: the use of trust funds (see *Trust and Money* report, 1993)

The report summarises what it considers are the fundamental principles of a modern contract. The points made are as follows:

- dealing fairly with each other in an atmosphere of mutual co-operation
- firm duties of teamwork, with shared financial motivation to pursue those objectives
- an interrelated package of documents, clearly identifying roles and duties
- comprehensible language with guidance notes
- separation of the roles of contract administrator, project or lead manager and adjudicator. The project manager should be clearly defined as the employer's representative
- allocating risks to the party best able to control them

- avoiding the need, wherever possible, of changes to pre-tender information
- assessing interim payments through milestones or activity schedules rather than through monthly measurement
- clearly setting out the periods for interim payments and automatically adding interest where these are not met
- provision of secure trust funds
- provision of speedy dispute resolution
- provision of incentives for exceptional performance
- making provision for advance payment to contractors and subcontractors for prefabricated off-site materials and components

Trust funds

It is fundamental to trust within the construction industry that those involved should be paid the correct amounts at the right time for the work that they have carried out. It may be argued that a problem does not exist and that

- employers only award work to firms with integrity
- contractors are at liberty to decline work from dubious employers
- subcontractors can adopt similar business practices
- bonds and indemnities are already available
- bad debts are not a problem unique to the construction industry

However diligently employers, contractors and subcontractors check each other out, the realities of the construction industry and its markets continue to exist. In circumstances such as a recession, contractors and subcontractors are prepared to undertake work for almost any employer. This is frequently done at a minimal profit margin. Bad debt insurance is available but adds extra costs at times when firms are seeking to reduce overheads. In times of prosperity, employers are prepared to undertake work with almost any firm available in order to get an important project constructed.

The construction industry is unique. Its characteristics separate it from all other industries. These include:

- the physical nature of the product
- manufacture normally takes place on the employer's own construction site
- projects often represent a bespoke design
- the arrangement of the industry whereby design is separate from manufacture is not mirrored in other industries
- the organisation of the construction process
- the methods and manner of price determination

The contractor's goods and services become part of the land ownership once incorporated within the project. Any retention of title clause that might be incorporated by suppliers or contractors in their trading agreements does not protect them once the materials are incorporated within the works. The contractor is also likely to be a considerable way down the queue, should an employer be unable to make payment within the terms of the contract. In some countries around the world, legislation has been provided to deal with the potential injustice that might be suffered. The most comprehensive is the Ontario Construction Lien Act 1993.

An effective way of dealing with this problem is to set up a trust fund for interim payments and retention monies. For example, an employer could be requested to pay into such a fund at the start of the payment period, e.g. at the beginning of the month. The correct payment, duly authorised, would then be paid to the contractor at the appropriate time. Where a form of stage payments was used, the amount of the particular programme stage would be deposited in the trust fund at the commencement of the work in this stage. Where a bill of quantities was used, the employer's quantity surveyor would request an appropriate amount of money to be placed in the trust fund to cover the next monthly certificate. The amounts authorised should correspond to the contractor's approved contract programme. The main contractor and the subcontractors would be informed of the amounts deposited. If any party considered the sums to be inadequate, they should have the option of applying to the adjudicator for their increase. There may be some argument for making payments to the subcontractors directly from this fund rather than through the main contractor's account.

Whether trust funds should be provided for all contracts is a matter of some debate. Some in the industry suggest that trust funds should only be available for projects with long contract periods or for contracts of a certain monetary size. The figure suggested for the latter is £0.25m. On very small projects agreement is often reached to make no payment until all the work is fully completed.

Any monetary interest accrued in the trust fund belongs to the employer. Where the fund necessitates bank charges, these costs need to be determined at the time of tender.

Trust funds are not really required for public works projects, since it is not likely that the bodies involved will become insolvent. However, such funds will be a source of reassurance for subcontractors, if the main contractor becomes insolvent during the course of the work. If trust funds are to used in the construction industry, they should be used in the public as well as the private sector.

Compulsory competitive tendering (CCT)

The government philosophy behind compulsory competitive tendering is that if market forces are allowed to operate, services can be provided with greater efficiency and at a lower cost. Government accounting and purchasing policies have made it clear that value for money and not the lowest price should be the aim. This was endorsed in *Competitiveness – Helping Business to*

Win (HMSO, 1994). CCT was intended to lead towards better managed, more innovative and more responsive services. However, some argue that if the Transfer of Undertakings Regulations apply, many of the opportunities for cost savings could be lost. The provision of publicly funded services through CCT has been growing around the world in recent years. It remains highly controversial in Australia and in the industrialised areas of Europe and the USA. It raises fundamental questions about competition and ownership in the provision of such services. Some of the more problematic issues in policy implementation are:

• fair and effective competition
• incentive compatibility
• performance monitoring
• whether CCT provides the best value for money

The preliminary assessments of contracting suggest generally successful outcomes. Empirical evidence at the present time indicates that efficiency gains have been made and effectiveness and quality of service have been maintained, if not enhanced. Few professional consultants, who come within this directive, are likely to admit openly that they have reduced their services because of professional fees. However, a survey by the Association of Consulting Engineers found that less time, resources and consideration was given to projects where competition on fees was applied.

Whilst the basic legal principles of CCT are also applied throughout the EU, through Council Directive 77/187, recent studies have indicated that as many as 99 per cent of public contracts are awarded within the same member state, for some of the following reasons:

• favour own national policies
• lack of information and advertising of contracts to tender
• discriminatory technical specifications and standards
• ignorance of legal procedures
• excessive use of exclusion rules or restrictive procedures

Reverse auction tendering

Under reverse auction tendering (RAT) contractors bid against each other in a live telephone auction to offer the lowest possible price or the best value for money bid. Bidders remain anonymous, with their bids relayed through an auction assistant to the auctioneer, who acts for the employer. The aim is to do away with the often unfair practice of one-off, sealed bids and to offer contractors a chance to lower their bids against their competitors. Whilst a number of different employer groups are considering piloting this idea, contractors are understandably much less enthusiastic. Contractors are raising issues of confidentiality, cartels, inequitable pricing and intellectual property rights. It is important that employers obtain the best possible price. However, if such a system resulted in contractors bidding too low to obtain work, this might have repercussions in terms of disputes arising or even more liquidation amongst construction firms and their suppliers. This would benefit no one.

Project partnering

During the 1980s the United Kingdom construction industry suffered one of its most severe recessions this century. Many of the largest and best contractors and design firms made losses in successive years and many of the smaller and not the worst firms ceased to trade.

There is a growing recognition in the construction industry of the need to move away from the confrontational relationships which cause the majority of disputes, problems, delays and ultimately expense.

The definition of partnering proposed by the National Economic Development Council is a long-term commitment between two or more organisations for the purpose of achieving specific business objectives by maximising the effectiveness of each participant's resources.

The Banwell Report (1967) recognised that there was scope under certain circumstances for the awarding of contracts without the use of competitive tendering – for example, where a contractor had established a good working relationship with an employer over a period of time, completing projects within the time allowed, at the quality expected and for a reasonable cost. These circumstances may have arisen through serial contracting and especially in the case of continuation contracts where projects have been awarded on a phase basis.

There has always been some reluctance on the part of public bodies to adopt such procedures, since they may lack elements of accountability, even though it could be shown that a good deal had been obtained for the public sector body concerned. Public sector bodies also need to follow EU procedures where appropriate. In 1995 a government white paper on purchasing gave the official seal of approval to partnering in the public sector. This provides an environment that is similar to that in the private sector, with a number of construction firms developing long-term relationships with public employers. Effectively, partnering will be possible where:

- it does not create an uncompetitive environment
- it does not create monopoly conditions
- the partnering arrangement is tested competitively
- it is established on clearly defined needs and objectives over a specified period of time
- the construction firm does not become over-dependent on the partnering arrangement

Partnering is now well established in the USA, offering benefits to all those who are involved. It has also been used successfully in the UK by a number of different major companies such as Marks and Spencer, Rover Group, British Airports Authority and Norwich Union. The partnering arrangement may last for a specific length of time or for an indefinite period. The parties agree to work together in a relationship of trust, to achieve specific objectives by maximising the effectiveness of each participant's resources and expertise. It is most effective on large construction projects or projects that are repeated throughout the country where the expertise developed can be retained and used again. The McDonald's chain of restaurants provides a good example of the latter. This arrangement, coupled with innovative construction techniques, has helped towards reducing the costs and time of construction by 60 per cent since the start of the 1990s. Its on-site activities have been reduced from 155 days to 15 days for a typical restaurant building.

The concept of partnering extends beyond the employer, contractor and consultants and includes subcontractors, suppliers and other specialist organisations which are able to add value to the project. The establishment of good relationships in the industry, based upon mutual trust, benefits the employer.

Appointment of specialist firms

The traditional arrangement of appointing specialist firms on a construction project is to use one of the nomination procedures (see Chapter 17). Whilst many specialist firms would like this procedure to be extended, it has been estimated that under JCT80 as few as 11 per cent of specialist engineering contractors are appointed in this way (Latham). Its decline in use is due in part to the long and tedious procurement route that has been advocated using this form. Alternative methods are available, such as:

Joint ventures: This is a particularly helpful approach where there is a large engineering services input to the project. The joint venture arrangement is between the main contractor, who may typically carry out the role of a project manager, and the specialist contractor. The companies work together as a joint company. It is therefore suitable for design and build arrangements.

Separate contracts: In this case the employer will let different contracts to different firms, i.e. the main contractor and the specialist contractor. This has not always been easy to administer, particularly where problems have arisen.

Management/Construction management: This is believed to be the most effective way for dealing with such firms. The different trade and specialist firms are appointed and a contractual arrangement formed with each company (see Chapter 7). This arrangement allows for full participation by the firms in design and commercial decisions at an early date.

Appointing a specialist firm as the main contractor: Where the specialist work represents the largest portion of the project, the employer may choose to reverse the arrangements and appoint a specialist firm as the main contractor. The more usual construction trades would then be employed by this firm.

Quality assurance

Every employer in the construction industry has the right to assume a standard of quality that has been specified for the project. The Building Research Establishment (BRE) has, in conjunction with a number of sponsors, recently launched the Construction Quality Forum. At the launch of this organisation, BRE reported that each year defects or failures in design and construction cost the industry and its employers more than £1bn per year. This represents 2 per cent of total turnover. This is a high percentage in view of the fact that contractors typically only make 6 per cent profit on turnover (R. C. Harvey and A. Ashworth, *The Construction Industry of Great Britain*, Butterworth-Heinemann, 2nd edn 1997).

The construction industry is an industry in which:

- There has never been a requirement for the work force to be formally qualified and skills are generally developed through time serving.
- Much of the work is carried out by subcontractors in a climate in which some 50 firms come into existence every day and a similar number go into liquidation or become bankrupt every day.
- There is a paucity of research and development involving new materials, designs and techniques.
- There is often poor management and supervision.

Studies have indicated that about 50 per cent of faults originate in the design office, 30 per cent on site and about 20 per cent in the manufacture of materials and components. An investment in quality assurance methods can therefore reap substantial long-term benefits by helping to reduce such faults, the inevitable delays and costs of repairs and the legal costs that all too frequently follow.

ISO 9000 certification has been increasingly taken up within the construction industry by consultants and contractors. Quality assurance is therefore seen as a good thing for the industry. Contract procurement methods that fail to address this issue adequately are not doing the industry or its employers any favours. The use of quality-certified firms, which have been independently assessed and registered, therefore offers some protection and a better chance of 'getting it right first time'. Work that is below acceptable quality and standards and has to be rectified is rarely as good as work carried out correctly in the first place.

ISO 9000 is seen by some firms as an additional expense, with the cost of accreditation as an unnecessary overhead. Also, whilst it should ensure that quality standards are achieved, it does not ensure that the appropriate quality has been set in the first place. However, some employers are no longer prepared to employ consultants, contractors or suppliers who do not have this kitemark. In the context of total quality management, quality remains a process of continued improvement. Quality must be appropriate to the work being performed. It should only be insisted upon where it adds value to the finished construction project.

Value for money

A major recommendation of the Latham Report is that (initial) construction costs should be reduced by 30 per cent. The report does not suggest how this might be achieved. Such cost reductions should not reduce quality or standards but should at least maintain and preferably improve these in construction projects. When employers compare new construction projects against other major capital investments such as plant and machinery, they score poorly against a number of different criteria. Some of their criticisms include:

- limited consideration of costs in use
- use of low technology
- lack of a comprehensive guarantee
- recurrent faults during the early months of use
- overall process too lengthy
- separation of design from construction

The implication for the construction industry is that of adding value, together with the principle of doing more with less. It is also important that such cost reductions do not refocus the emphasis within the construction industry on initial costs alone, as was the case fifty years ago. The importance of ensuring that life cycle costs are given their rightful importance in the overall construction process must be maintained.

The reduction of construction costs must be considered in the light of the fact that the industry is currently in one of its worst recessions this century. It is a low wage industry and cost efficiencies have been achieved over the past fifty years principally through the study of construction economics. The following are some of the areas of possible investigation in attempting to meet this aim:

Macro issues
- Apportion risk efficiently
- Improve productivity
- Reduce waste
- Examine cost-efficient procurement arrangements
- Improve the use of high technology for both design and construction
- Reduce government stop-go policies
- Develop more off-site prefabrication of components

- Standardise more components
- Consider construction as a manufacturing process
- Improve the education and training of operatives and professionals
- Reduction of the size of workforce, noting their high costs and their reduction in manufacturing industry

Micro issues
- Reduce the need for changes to the design
- Optimise specifications
- Improve design cost-effectiveness
- Get it right first time, i.e. avoid defects
- Make better use of mechanisation

Private finance initiative (PFI)

The purpose of this initiative was to encourage partnerships between the public and private sectors in the provision of public services. The scheme is outlined in *Private Finance and Public Money* (Department of the Environment, 1993). In 1992, the Chancellor of the Exchequer announced a new initiative to find ways of mobilising the private sector to meet needs that had traditionally been met by the public sector. Achieving an increase in private sector investment will mean that more projects will be able to be undertaken. This takes into account the government's objective that public spending should decline in the medium term. The broad aims of such a partnership will be to:

- Achieve objectives and deliver outputs effectively
- Use public money to best effect
- Respond positively to private sector ideas

In exploring the possibilities for private finance, including proposals from the private sector, the questions being considered include:

- Can the project be financially free-standing?
- Is it suitable for a joint venture?
- Is there potential for leasing agreements?
- Is there potential for government to buy a service from the private sector?
- Can two or more of these elements be brought together in combination in any particular instance to form innovative solutions?

Concessionary contracting falls neatly into such an arrangement, whereby the private sector is encouraged to construct public projects, such as roads, and then charge a levy on this provision for a fixed period of time specified in the contract. The contractor is responsible throughout the entire period for the maintenance of the works. The contract will also specify the required condition of the asset upon eventual handover to the public sector.

PART THREE

ORGANISATION AND PROCEDURES

CHAPTER 10

PROCESS AND PARTIES OF THE CONSTRUCTION INDUSTRY

Introduction

Civil engineering projects can be separated into two stages, pre-contract and post-contract activities. This separates the project into design work and construction work. Projects can also be identified in four separate stages or phases by separating the tentative (feasibility) from the definite (design) and the construction from the maintenance. The last of these is a much under-rated aspect of construction activity. When budgets have to be reduced within organisations, it is often the repairs and maintenance budget that suffers the worst. However, in terms of its relative time span and importance, and the need to protect an investment its significance has increased considerably in recent years.

Pre-contract process

This is the period during which the need for the project, and the idea in terms of size, function and appearance, are formulated into plans that are capable of being used for the construction of a complete structure. Changes to the design will continue to occur throughout this period. Variations are also expected even after the work has commenced on site, but this should not be seen as an excuse for a delay in decisions in terms of the design and specification requirements. Indications are currently showing that this lack of decision prior to commencement on site is an important factor affecting the important trio of time, cost and quality of the finished project.

Employer's brief

On a traditional project the employer will first engage an engineer to work on the outline ideas for the project. In some circumstances the employer may decide to go straight to the contractor for a design and construct service, or to a firm who can offer an 'off-peg' project that may meet the requirements identified in the brief. It is much more common to involve the contractor in the pre-contract design on building projects than it is on civil engineering projects. The purpose of this first meeting between the employer and the engineer is to distinguish between what is needed and what is desired. Many employers will have a good idea of what is required and how much money they have to complete the work. The two are often incompatible! Good up-to-date cost information is essential at this stage if the project is to be successful. The money available may be determined on the basis of the employer's capital resources and funding ability. Possible alternative design solutions will need to be considered, along with the appropriate construction costs.

Investigation

Assuming that the initial requirements can be agreed, the project moves towards a second stage. During this period, many of the aspects are examined to ascertain the viability of the project. A site survey and investigation should be carried out to determine the nature of the ground and site conditions, and, where possible, to locate the proposed project in the advantageous position on the site. This advantageous position will consider the design restraints that may be present and the necessity to achieve an economic solution and value for money. Bad ground conditions may have implications for the overall aspects of the design. This will inevitably increase construction costs. Alternative methods of construction, both in terms of design and costs, will be considered. Once the outline design is approved, the scheme should be cost planned to ensure a balanced design that meets the overall requirements of the design brief.

Sketch design

During this stage the engineer will be able to obtain outline planning approval and meet other statutory requirements regarding the project. There will previously have been discussions with the planners on the possibility of such a scheme being approved. Where necessary, the engineer will have consulted the structure and district plans to ensure at this early stage that the proposed development will not contravene these plans. The major planning problems will thus be solved.

Design

This is the period during which the sketch plans are developed and the constructional details and methods of construction are determined where possible. Unlike building projects, where the project often determines the constructional methods to be used, civil engineering projects are heavily dependant upon a particular contractor's method of working. Where possible, the cost implications of the various solutions should be assessed to confirm that the cost remains on target.

Working drawings

All of the drawn information should be completed during this stage. The practice of leaving parts of the design until later is to be deprecated. If the scheme has been fully designed, it should now be possible to obtain quotations from nominated subcontractors. Prior to commencing the preparation of the bill of quantities, the costs of the design should be checked against the latest cost plan. In this way, when the tender sums are submitted they should show some comparisons with the latest estimated forecasts of costs.

Tender stage

During this stage of the project, tenders are invited from firms who have already expressed their willingness to submit a price. While tenderers are busy pricing the documents, the engineer and the other consultants will be checking through their own calculations and designs. However, the pressure for progress has now moved almost entirely to the contractor. During pricing, the latter will obtain domestic subcontractor quotations and current information from various suppliers on the costs of materials, components and goods. The contractor will also need to be aware of the current state of the market in order to submit the most favourable price. Tenders will be returned by a stated date in the stated way. Finally, a tender will be recommended for acceptance and, once the contract has been signed, the project moves into the post-contract phase.

Post-contract process

Construction

The project should have been completely designed prior to release to the contractor for tendering purposes. Although the contract allows for and expects variations to occur, this should not provide an excuse for a design which is only partly finished. One of the engineer's main duties during this stage is to ensure that the contractor has all the information required for construction purposes. In addition, the engineer will be responsible for the smooth running of the works. As the work is carried out, the engineer's role is largely that of a supervisor, ensuring that the contractor complies with all the requirements. During the construction of the works, the quantity surveyor will prepare the valuations for the interim certificate. Certification is, however, entirely the responsibility of the engineer. It is preferable if the works can be remeasured and agreed soon afer they have been completed. In this case the work items will still be clear in everyone's minds, and the possibility of errors arising is therefore reduced. The various subcontractors will come onto the site to carry out their own work. Even the major subcontractors will only be on site for a relatively short period of time, compared with the main contractor. The engineer is unlikely to have completed the process of nomination for every firm, and some of this will therefore need to be done during the construction stage. The local authority will have little to do with a civil engineering project in the majority of circumstances, but may choose to inspect the works at different stages. This will be done to make sure that each stage complies with the appropriate regulations and bye-laws. The local authority will, however, take action where local regulations are infringed or where a nuisance is caused to nearby properties due to the carrying out of the construction works. The Health and Safety executive may visit the site to check that the works are being carried out in accordance with good practice.

Maintenance

The project becomes officially complete when the engineer issues the certificate of substantial completion of the works. The contractor is responsible for making good any of his defects from this date for about 6 months or whatever period of time is stated in the Appendix to the conditions of contract. The engineer should ensure that all defects are made good prior to the issue of the final certificate. The quantity surveyor during this period will prepare and agree the final account with the contractor's surveyor or measurement engineer. At the commencement of the defects liability period, one-half of the retention is released to the contractor, with the other half being paid with the final certificate. Although the contractor is contractually responsible for work up to the end of this period, liability under common law will extend for a much longer period.

Parties involved with the construction process

The employer

The main purpose of construction activity is to provide a completed project for the owner or promoter. This project may include traditional contract work, a project constructed speculatively, in-house work using direct labour, or infrastructure works such as roads, bridges or pipelines. The employers, clients or promoters of the construction industry are many and varied. They may have virtually any legal status. They include the following.

Central government

Contracts are arranged through a government department, such as the Department of the Environment. Funding for projects is provided by a vote of Parliament. Any expenditure in excess of this vote is sanctioned by the department, although subject to a supplementary vote. The work that each department is allowed to commission is determined by the government in power.

Local government

Local authorities act under Charters and Acts governing their procedures, the type of work that they are allowed to carry out and the methods by which they can raise funds for capital works projects. They are subject to the ordinary legal liabilities as to their legal powers to contract and their liability to be sued.

National or public industries

These have reduced considerably in number during the past decade, largely because of the transfer of their assets to the private sector. The legal position of such bodies and of their officials is similar to those of local authorities.

Public liability companies (PLC)

These now include the utilities of water, gas and electricity. These contract under their own terms and arrangements in a similar manner to other private companies. They are answerable to the respective government regulator in ensuring that they secure value for money in all their activities.

Employers generally

Each separate organisation is granted a considerable amount of autonomy by the government of the day, and there is a wide diversity of methods used for the procurement and execution of major and minor capital works projects. Little uniformity exists either in the design procedures employed or the contract conditions used. Whilst the private sector comprises all those organisations that are not a part of a government organisation or agency, they may nevertheless be able to secure a contribution from government as an incentive to complete their construction works.

Employers in the public sector may be influenced by both social and political trends and needs, and the desire to build may be limited by these factors, though the principal restriction on their aspirations is the amount of capital they are allowed to borrow. The private sector, from the individual house-owner to the large multinational corporation, generally directs its capital spending to the ventures that result in monetary or social benefits. In recent years there has been an increased emphasis upon securing value for money. This has tended to be viewed over the project's life cycle rather than judged on its initial construction costs alone.

The employer is one of the parties to the contract, the other being the contractor. Each employer (Table 10.1) will have different priorities but essentially these will be a combination of the following:

- **Performance** in terms of quality, function and durability
- **Time** available for completion by the date agreed in the contract documents

Table 10.1 Employer requirements

Performance	Time	Cost
Aesthetics	Design period length	Initial budget
Quality	Start on site date	Approximate estimate
Function	Hand-over date	Tender sum
Durability	Final completion	Final cost
Maintenance		Cost in use

- **Cost** as determined in the budget estimate and the contract sum. If employers are to be satisfied with their construction project, then these three conditions must always be critically considered

The employer's main responsibilities regarding the project are to:

- define the extent of the project and the functions it is to perform
- provide information required by the engineer during design and construction
- obtain the necessary legal authority to allow the construction of the project
- secure funding for the project and ensure that this is available at the appropriate times
- acquire the necessary land for development

There are several references to the employer in the ICE Conditions of Contract. These include the following:

- definition: 'Employer means the persons or persons, firm, company or other body named in the Appendix to the Form of Tender and includes the employer's personal representatives, successors and permitted assigns.' (clause 1 (a))
- indemnity for third party exceptions (clause 22)
- indemnity for unavoidable noise disturbance and pollution (clause 29)
- responsibility for safety when using own workmen (clause 19)
- licensee under the New Roads and Street Works Act 1991 (clause 27)
- bearing the expense for excepted risks (clause 20)
- giving details when certificates and payments differ (clause 60)
- giving timely access to the site (clause 42)

General contractor

The majority of the construction work in the United Kingdom is undertaken by a general contractor. These firms, which will be public limited companies (PLCs), will vary in size, having from just a few to many hundreds of employees. Many of the larger companies are household names and have developed only since the beginning of the twentieth century. About a dozen of these firms have turnovers in excess of £1bn. The top 50 construction firms account for about two-thirds of all contractors' workloads within the construction industry. The former Federation of Civil Engineering Contractors included about 300–350 civil engineering contractors.

Although there is no clear dividing line between civil engineering and building works, many firms tend to specialise in only one of these sectors. In the larger companies, separate divisions or companies exist, often trading and structured in entirely different ways, depending upon the sector in which they are employed. For example, building companies tend to be organised on a regional basis, whereas civil engineering companies will operate nationally. Even the operatives' unions and the rules under which they are engaged are different. The smallest firms may specialise in a single trade, and may act as either domestic or nominated subcontractors The largest firms may be almost autonomous units, although it is uncommon even in these companies to find them undertaking a complete range of work. On the very large projects it is usual to find specialist firms for piling, steelwork, etc.

Contractors under the ICE Conditions of Contract, and most other forms of contract, agree to carry out the works in accordance with the contract documents and the instructions from the engineer (architect under JCT). They agree to do this usually within a stipulated period of time and for an agreed amount of money. The main contractor must also comply with all statutory laws and regulations during the execution of the work, and ensure that all who are employed on the site abide by these conditions. The contractor will still be responsible contractually for any defects that may occur, for the period of time stipulated in the conditions of contract (typically 6–12 months). However, the responsibility of the contractor for the project does not end here. In common law the rights of the employer will last for 6 years or 12 years, depending upon whether the contract was under hand or seal. The contractor is mentioned extensively in the conditions of contract, largely because, along with the employer, the contractor is one of the parties to the contract. Some of the more important provisions include:

- definition: 'Contractor means the person or persons, firm or company to whom the contract has been awarded by the employer and includes the contractor's personal representatives successors and permitted assigns.' (clause 1(b))
- responsibility for work of subcontractors (clauses 4, 59)
- permanent works designed by contractor (clause 7)
- to provide further documents (clause 7)
- responsibility unaffected by engineer's approval or consent (clauses 7, 14)
- general responsibilities (clause 8)
- design responsibility (clause 8)
- responsibility for safety (clauses 8, 19)
- work to be as contract and to engineer's satisfaction (clause 13)
- to provide programme and revisions (clause 14)
- indemnity to employer (clause 24)
- employer's remedy on failure to insure (clause 25)

- to conform with statutes (clause 26)
- giving notices under the NRSWA 1991 (clause 27)
- to pay all royalties (clause 28)
- to take precautions against damage to highways, etc. (clause 30)
- to bear costs of additional access and facilities (clause 42)
- liquidated damages for delay (clause 47)
- to search for cause of defect, etc. (clause 50)
- to attend for measurement (clause 56)
- responsible for nominated subcontractors (clause 59)
- default by contractor (clauses 39, 62, 63)
- abandonment of contract, bankruptcy, etc. (clause 63)
- appointment as principal contractor (clause 71)

Contractor's agent

The contractor's agent is responsible for the effective control of the contractor's work and workpeople on site. The agent is also responsible for organisation and supervision on the contractor's behalf, and for receiving instructions from the engineer. Depending upon the size and nature of the works and the type of firm, the agent's role may range from that of a general foreman to that of a project manager. The agent may have received initial training as a trade craftsman and then became foreman or ganger, or be a chartered engineer. The responsibilities will vary with the size of the project and company policy. On the larger projects considerable assistance will be received from other site staff.

Clause 15 of the ICE Conditions of Contract requires the contractor to provide all necessary superintendence during the construction and completion of the works. This may also be required after completion, at least whilst any defects are rectified under the terms of the contract. The superintendence provided must be by someone who has an adequate knowledge and experience of the sort of works being contemplated. This will include knowledge of the methods and techniques to be used and the hazards that might be encountered. Aspects of site safety should be fully considered and methods and procedures adopted that are likely to prevent accidents from occurring.

The agent must be competent and authorised by the contractor and approved of in writing by the engineer. The agent must be available on site for the whole time in order to supervise the works properly. The agent will receive directions from the contractor (the agent's employer) and instructions from the engineer or the engineer's representative on behalf of the employer (or promoter).

Structural engineers

Structural engineering is really a branch of civil engineering concerned with the analysis of structural capability. Structural engineers ensure that a building or other structure is able to withstand the different loads and pressures which might be applied. These include the dead loads such as the weight and forces of the structure itself; applied loads such as people, equipment, machinery (and the vibrations which they might cause); environmental factors such as wind pressures; the pressures caused through thermal expansion and contraction and even forces caused by earth tremors. Structural engineering is defined as the science and art of designing and making with economy and elegance buildings, bridges, frameworks and other structures, so that they can safely resist the forces to which they might be subjected.

The civil engineer

There is a wide range of different types of engineers employed both within and outside the construction industry. Civil engineers are responsible for the design and supervision of civil and public works engineering, and are employed in a similar way to architects on building contracts. The civil engineer's powers and responsibilities under the ICE Conditions of Contract are wide ranging. The engineer's counterpart working for the contractor is also likely to be a civil engineer. The work of the civil engineer is diverse, and may include projects associated with transport, energy requirements, sewage schemes or land reclamation, typically costing several millions of pounds. The demarcation between civil engineering and building is ill-defined and there is an overlap, with some projects falling into each area. To give some idea of their size and complexity, typical civil engineering projects include: the Humber Bridge, Thames Barrier, Docklands Light Railway, Sizewell Nuclear Power Station, Electrification of East Coast Mainline and the Channel Tunnel project. The latter was one of the largest projects ever undertaken and was a joint venture with French civil engineers and contractors.

Advice regarding suitable consulting engineers and their methods of engagement and working can be obtained from the Association of Consulting Engineers. The selection of an appropriate consulting engineer should start with an assessment of the qualifications and experience of possible firms or individuals. Corporate membership of the Institution of Civil Engineers is usually a prerequisite. This will ensure that appropriate education, training and experience will have been undertaken. Final selection is usually through inviting proposals, often on the basis of fee competition.

When an employer engages a firm of consulting engineers, a formal agreement should be completed which sets out the duties and responsibilities involved and the fees and expenses that are to be paid. The Association of Consulting Engineers 'Conditions of Engagement' provide clear guidance on the various matters that should be covered.

The duties of the engineer will include the development, design and technical direction of the works. This will include the preparation of drawings, specifications, bills of quantities and other documents. It is common today for the engineer to engage non-engineering specialists from other professions to assist with some of this work. A firm of consulting engineers may therefore include economists and planners (to carry out economic and social investigations), geologists (for site investigation works), quantity surveyors (for bill preparation and remeasurement) and architects (for aspects of aesthetic design).

References to the engineer in the ICE Conditions of Contract include the following:

- definition: 'Engineer means the person, firm or company appointed by the employer to act as engineer for the purposes of the contract and named in the Appendix to the Form of Tender or such other person, firm or company so appointed from time to time by the employer and notified in writing as such to the contractor.' (clause 1 (c))
- delegation of powers to engineer's representative (clause 2)
- duties and authority (clause 2)
- named individual to act (clause 2)
- to act impartially (clause 2)
- instructions in writing (clause 2)
- appointment as planning supervisor (clause 7)
- acceptance or rejection of the contractor's programme (clause 14)
- consent to contractor's methods of construction (clause 14)
- failure to disapprove work or materials (clause 39)
- to fix rates for variations not covered by the contract (clause 52)
- action on contractor's objection to or determination of nominated subcontract (clause 59)
- power to correct or withhold certificates (clause 60)
- decision on dispute (clause 66)
- witness in arbitration (clause 66)
- certificates net of Value Added Tax (clause 70)

Engineer's representative

Clause 2 (3) of the ICE Conditions of Contract states that the engineer's representative is responsible to the engineer. The contractor must be informed in writing who can represent the engineer. The duty of the engineer's representative is to watch and supervise the construction up to the completion of the works. The representative has no responsibility to relieve the contractor of duties or obligations. Unless the contract expressly provides, the representative cannot order any work that might result in extra costs to the employer, nor make variations. Whilst a representative may issue instructions to the contractor these have no force unless they are subsequently authorised by the engineer. The contractor must give the engineer's representative every reasonable facility to carry out all duties. These include ensuring that the contract is fully complied with in terms of the specification and further instructions from the engineer. The representative will attempt to make sure that the materials used and the workmanship carried out are in accordance with the contract requirements. This will involve inspecting the materials prior to their incorporation within the works; obtaining samples where necessary for the approval of the engineer; testing materials such as concrete, bricks and timber to the specified British Standards and codes of practice and generally ensuring that the construction work complies with accepted good practice.

The quantity surveyor

The function of the quantity surveyor has developed from that of a measurer to one of a construction cost adviser. It is unusual today on either major civil engineering or building projects not to utilise the services of a quantity surveyor. For most of this century civil engineering contractors have seen the benefits to be gained by employing quantity surveyors to measure and value the works and to prepare contractual claims. The emphasis of the quantity surveyor's work has moved from one solely associated with accounting functions, to one involved in all matters of forecasting finance and costing. The function of the quantity surveyor in connection with construction projects is threefold:

- cost adviser, forecasting and evaluating the design in economic terms, both on an initial and life-cycle cost basis
- preparation of tendering documentation to be used by contractors
- financial accounting during the construction period; preparation of interim payments and financial progress; preparation, control and agreement of the final expenditure for the project, including the settlement of contractual claims

The architect

The architect is the civil engineer's traditional counterpart on a typical building project. In the building process, where design and construction are separate entities, it is the architect who has traditionally received the commission from the employer. The architect's function is to provide a proper arrangement of space within the building, its shape, form, type of construction and materials used, environmental controls and aesthetic considerations.

Subcontractors

It is very unusual today for individual contractors to undertake all the contract work using their own workforce. Even in the case of minor building projects the main contractor is likely to require the assistance of some specialist trade firms. Firms undertaking work other than the main contractor are often described as subcontractors, although in some situations it is not uncommon to find specialist firms working on the site beyond the normal jurisdiction and confines of the main contractor. The employer may, for example, choose to employ such firms directly, and in this context these firms are not to be considered as subcontractors of the main contractor. Provision is made in the conditions of contract for such firms to have access to the contractor's site.

Nominated subcontractors

The employer may choose to nominate particular firms to undertake the specialist work that will be required. This approach may be adopted in order to gain a greater measure of control over those who carry out the work. These subcontractors enjoy a special relationship with the employer (see clauses 58–59 in Chapter 17). Although after nomination they are often supposedly treated like one of the main contractor's own subcontractors, they do have some special rights, for example, in respect of their payment. The engineer may also choose to name in the bills of quantities or specification subcontractors who will be acceptable firms for the execution of some of the contractor's own work. This procedure avoids the lengthy process of nomination, but still provides a substantial measure of control on the part of the engineer. A number of approved firms are often listed to which a contractor may add further names for the approval of the engineer. Such approval should not unreasonably be refused. This provides for some measure of competition and choice of firm.

Domestic subcontractors

All the remaining work is still unlikely be carried out by the main contractor, and provision is also made for the use of the contractor's own subcontractors. These subcontractors are often referred to as the main contractor's domestic subcontractors (clause 4). The contractor:

- must not subcontract the whole of the works without the prior written consent of the employer
- can employ labour-only subcontractors without notifying the engineer
- remains responsible for the work of subcontractors
- after receiving a warning from the engineer, may be requested to remove from the site subcontractors who in the opinion of the employer are guilty of:
  misconduct
  incompetence
  negligence
  failing to conform on matters of safety
  persisting in any conduct which is prejudicial to safety or health

Neither the employer or the contractor can assign the contract in part or in whole without the written consent of the other party (clause 3). This consent should not be unreasonably withheld.

Suppliers

Construction materials delivered to a site can be described under several different headings, such as materials, components and goods.

Materials are the raw materials to be used for construction purposes and include, for example, cement, aggregates, bricks, timber, etc. The items included within this description will, in total, often represent the largest expense on the typical construction site. However, as more and more of the construction processes are manufactured off-site for on-site assembly, so the value of this section will diminish.

Components represent those items delivered to site in almost 'kit' form. They may include, for example, precast concrete beams and prefabricated steelwork. The industrialisation of the construction process is based to a large extent on the assumption that many items can be delivered to site in component form to be assembled easily on site. Whilst the costs of off-site manufacture may in some cases be more expensive, quality control and speed of construction are often much improved. Components are usually manufactured specifically for a particular project.

Goods include items that are generally of a standard nature, which can be purchased directly from a catalogue, for example, sanitary ware, ironmongery, electrical fittings. The contractor's source of supply for these items may vary and must comply with specifications in the contract documents regarding quality and performance. Some of the items will need to be obtained directly from the manufacturer. In other circumstances, specialist local suppliers of timber or ready-mixed concrete will be used.

Contractors are able to secure trade discounts for the items that they purchase, and such discounts may be increased either to attract trade or because of large orders. Some employers, who undertake an extensive amount of construction work, are often able to arrange a bulk purchase agreement with suppliers. Bulk purchase agreements result in lower costs for materials, goods and components to the employer. This helps to reduce the overall costs of construction. The contractor must then obtain the relevant items from the bulk agreement suppliers.

The ICE Conditions of Contract do not make any special reference to suppliers. The clauses (58–59, Chapter 17) of the conditions relate to subcontractors. The definition of nominated subcontractor (clause 1 (m)) indicates that such firms may be those who supply and fix materials or who supply the materials only, which will then be fixed on site by the contractor. The specification may also identify particular suppliers from whom the contractor must purchase materials, components or goods. It is usual to suggest a list of alternative suppliers or sometimes to add the words 'other equal and approved'. The contractor would then need to show that the items proposed to be purchased for the work complied with specifications.

The professional bodies in the construction industry

Designing, costing, forecasting, planning, organising, motivating, controlling and co-ordinating are some of the roles of the professions involved in managing construction, whether it be new build, refurbishment or maintenance. These activities also include research, development, innovation and improving standards and performance.

The Institution of Civil Engineers (ICE)

The term 'civil engineer' appeared for the first time in the minutes of the Society of Civil Engineers, which was founded in 1771. It marked the recognition of a new profession in Britain as distinct from the much older profession of military engineer. The members of the Society of Civil Engineers were developing the technology of the industrial revolution. There was no formal education even by the nineteenth century. In France, by contrast, the government had established the Grandes Ecoles to train engineers for the civil service. In order to remedy this situation in Britain, a group of aspiring engineers founded the Institution of Civil Engineers (ICE) in 1818. The most famous engineer of the day, Thomas Telford became its first president and in 1928 it was formally recognised by the granting of a Royal Charter. This charter contains the often quoted definition of civil engineering as being 'the art of directing the great sources of power in nature for use and convenience of mankind'.

The engineering committee of the Institution provides a forum in which engineers in different specialisms can exchange views and information. The designated areas include: ground engineering, water engineering, structural engineering, building technology, maritime engineering, transport engineering and energy.

The Institution of Municipal Engineers merged with the ICE in 1979. The engineering profession is very diverse, with a variety of engineering institutions coming under the umbrella of the Engineering Council. Altogether there are about 300,000 members. In addition, the Association of Consulting Engineers, acting as a voice for those in practice, has a membership which includes gas, chemical, electrical and mechanical as well as civil and structural engineers. Civil engineers work in the four main areas of contracting, consultancy, government and nationalised and denationalised industries. Consultancy accounts for the biggest proportion of members in employment, followed closely by those who work for contractors.

The Institution of Structural Engineers (I Struct E)

The Institution of Structural Engineers (I Struct E) began its life as the Concrete Institute in 1908, was renamed in 1922 and was incorporated by Royal Charter in 1934. Its aims include promoting the science and art of structural engineering in all its forms and furthering the education, training and competencies of its members. The science of structural engineering is the technical justification in terms of strength, safety, durability and serviceability of buildings and other structures. The majority of structural engineers are employed in private practice.

Association of Consulting Engineers (ACE)

This was formed in 1913 to help to promote the advancement of consulting engineering. It includes not only civil engineering but also branches of other types of engineering such as electrical and mechanical engineering. It seeks to ensure that engineers advising employers or clients are appropriately qualified and are bound by a code of professional rules and practice. The Association provides a list of engineering firms who are not directly or indirectly connected with any commercial or manufacturing interest which might influence their professional judgement. It liaises with government and other public organisations as well as the other professional bodies. It recommends rules and conditions of engagement which are accepted as a basic standards for the engineering profession.

The Association is a member of the Fédération Internationale des Ingénieurs-Conseils. This is an organisation of consulting engineers' associations in a number of countries throughout the world which have similar interests and activities.

The Royal Institution of Chartered Surveyors (RICS)

The Institution was formed in 1868 and incorporated by Royal Charter in 1881. It originated from the Surveyors' Institute and has grown as a result of a number of mergers with other institutes, most notably the Land Agents (1970) and the Institute of Quantity Surveyors (1982). The RICS is currently administered in seven divisions which represent the varying interests of different chartered surveyors. The General Practice (42 per cent) and Quantity Surveying (37 per cent) divisions account for over three-quarters of the membership.

The Royal Institute of British Architects (RIBA)

The RIBA is the main professional body for architects in England and Wales. In Scotland, there is a similar body, the Royal Incorporation of Architects in Scotland (RIAS). Under the Architects Registration Act 1938 it continues to be illegal for anyone to carry out a business describing themselves as an architect unless they are registered with the Architects' Registration Council (ARCUK), established under an Act of 1931. Registration involves appropriate training and education, as evidenced by the possession of qualifications set out in the Act or approved by the council. However, this does not prohibit anyone from carrying out architectural work such as the design of buildings.

The Chartered Institute of Building (The CIOB)

The CIOB was formed in 1834, incorporated under the Companies Acts in 1984 and granted a Royal Charter in 1980. The objectives of the Institute are the promotion, for public benefit, of the science and practice of building, the advancement of education and science, including research, and the establishment and maintenance of appropriate standards of competence and conduct for those engaged in building. The Institute encourages the professional manager and technologist to work together with their technician counterparts in order to achieve a ladder of opportunity as a main objective of the training and examination structure of the Institute.

Chartered body membership

Table 10.2 lists the main professional bodies working in the construction industry and the relative membership sizes.

There are also a number of non-chartered professional bodies with memberships ranging from as few as 1,000 to nearly 10,000. These include the Institution of Civil Engineering Surveyors (ICES) and institutions from other countries such as the Association of Cost Engineers which has strong links in the USA. Other professional bodies have been formed as a result of the changing needs and emphasis of the construction industry, for instance, the Association of Project Managers.

Table 10.2 Professional body membership

Chartered institution	Membership	
	Chartered	Total
Royal Institution of Chartered Surveyors	71681	93256
Institution of Civil Engineers	49726	79606
Chartered Institute of Building	12145	32370
Royal Institute of British Architects	27227	31700
Institution of Structural Engineers	12513	22825
Royal Town Planning Institute	13355	17726
Chartered Institution of Building Services Engineers	7495	15363

Construction employers' associations (CEC)

The construction industry has typically been seen as two sectors, building and civil engineering. This distinction was also seen at employer level with the separate Building Employers' Confederation (BEC) and the Federation of Civil Engineering Contractors (FCEC). In 1997 BEC and FCEC agreed to a merger, becoming the Construction Employers' Confederation (CEC). This organisation seeks to speak with a single voice for the contractors' interests at national level and to influence government policy for the benefit of its members. Most of the largest 20 firms belonged to the former BEC or FCEC. BEC had a membership of over 10,000, latterly this fell to 8,000. Prior to the recession of the early 1990s, the number of firms in FCEC was much smaller at 300 (700 previously), although the typical size of firm was much larger. In addition, there remain other subsidiary groupings of firms such as the House Builders' Federation and committees for special interest groups. In recent years, there has been an increased awareness of matters such as European harmonisation, health and safety and reducing the costs of construction while retaining value.

Many of the contractors believe that the construction industry has a poor image, a divisive educational framework and needs to improve relationships with employers and the professions. The single voice should allow for improvement in the lobbying strength to government for investment in the industry.

Construction Industry Training Board (CITB)

The Construction Industry Training Board was established by an Act of Parliament in 1964 to help provide enough trained people for the construction industry by improving the quality of training, and the facilities available for training. It is partially funded through government, but most of its income is derived from a levy system on contractors, based upon the number of their employees. In addition to being appointed as the primary managing agent for the construction industry's Youth Training Scheme, the CITB also provides for a range of skills training at its national training centres. Whilst the emphasis of its activities is on practical craft skills training, it also offers courses in supervisory management.

CHAPTER 11

ORGANISATION AND PLANNING

Introduction

The construction process includes all activities involved in bringing to fruition a construction project from its conception to its physical completion on the site. The planning of construction works is fraught with difficulties. Even a relatively small project needs careful planning in order to try to ensure a balance between progress, quality, safety and economy. Although planning will not solve the problems of inclement weather, delay in the delivery of materials and plant breakdowns, it will provide a comparison of the work planned against that actually achieved.

The construction industry is usually involved in one-off projects; each of these is invariably managed with a new team. As the location of each project varies widely, the workforce is largely new and the conditions under which the project is undertaken can differ, depending on the site conditions, climate, etc. The size of the main contractor's site organisation, which is comprised of technical and non-technical staff, often depends on the size of the works being constructed. However, as the use of sub-contractors has become more popular, the main contractor's team has reduced in size. In the case of a management contractor taking responsibility only for the construction of the project, all of the site work is subcontracted.

Once the design of the project is completed it is the responsibility of the contractor to construct and maintain the project in accordance with the contract documents. The prime requirements are fitness for purpose and the maintenance of safety during construction and operation. In terms of the work on site, the essential task is to construct the works in accordance with the drawings and specification. The contractor is given a measure of freedom to carry out the project in a way that is appropriate, including the design and execution of temporary works. The ingenuity of the contractor in providing inexpensive temporary works may reflect a keen price for a project but may also represent optimism and cutting corners. Although the designer might check the contractor's proposals, this in no way relieves main contractors of their responsibilities and liabilities. The contractor should welcome an independent checking system in the interests of the safety of personnel and the works.

Table 11.1 Planning and control systems

Objectives	Planning systems	Control systems
Progress	Programming	Progress control
Quality	Drawings	Quality control
	Specification	
Safety	Law on safety	Safety statistics
	Contract clauses	
	Site instructions	
Economy	Estimating	Cost control
	Budgeting	Budgetary control
	Cash flow forecasting	Cash flow control
	Profit forecasts	Financial control

For each objective there is an established planning and control system. These are shown in Table 11.1, which has been adapted from the *Civil Engineer's Reference Book*.

The construction industry has many examples of projects where planning was weak, i.e. deciding what is wanted and how it is to be achieved, or where control was ineffective, i.e. comparing what has been done with what ought to have been done.

Contractor's organisation

Once a contractor's tender has been accepted, the contractor will need to consider establishing a management organisation on site. The organisation will vary depending upon the nature of the works being constructed and the size of the project. The nature of the site organisation will also depend upon how the firm itself is organised in terms of the relationship between head office and the site.

A civil engineering company can be arranged and organised in many different ways. This will depend to some extent upon the company objectives and the strategy used to achieve these objectives. Fundamental to any organisation is team work and recognising that all departments and individuals have a part to play in the success of the company. The company objectives will need to consider the kind of services the company can offer and the demand in the market for those services. It will need to decide on what share of the market it requires, and be constantly examining indicators such as workloads, in order to maximise its potential. When the market is in decline it will need to consider either reducing its own operations or seeking out alternative sources of work.

The contractor must provide the following services in some way:

Engineering management: Design, supervision, quality control, work study, research and development

Commercial management: Marketing, finance, estimating, tendering, quantity surveying, legal

Operations management: Plant and transport, workshops, operatives

Administrative management: Personnel, salaries, purchasing, health and safety, education and training

Contractor's site organisation

A general arrangement for a civil engineering project is shown in Figure 11.1. An agent or project manager is appointed to manage the particular project. Civil engineering projects are frequently of a size to require an agent who is resident on the site. The agent is normally a chartered civil engineer, usually with experience of the type of work that is envisaged. In addition to engineering knowledge and skills, the agent must also possess qualities of leadership and management. The main duties include ensuring that the works are carried out in accordance with the contract and that the contractor's work is completed as economically as possible to the satisfaction of the engineer. On a large project it is frequently necessary to employ sub-agents to take responsibility for parts of the project or for some of the activities that need to be carried out.

Essentially, the organisation can be subdivided into the following groups. They often mirror the organisation at the head office of the firm.

Engineering management

On a large project a number of site engineers will report to a sub-agent on particular aspects of the work. For example, on a highway project it is usual to subdivide the work into sections. This may be a division between roads and bridgeworks or for particular stretches of road. They will be responsible for the accurate setting out of the work, for the control of the construction and for its eventual quality. In their work they will be supported by geologists and scientists, who will be responsible for testing and analysing the different materials. The contractor's site engineers are responsible for implementing the contract and instructions from the employer's engineer.

Commercial management

The commercial management activities on site will be carried out by measurement engineers or quantity surveyors. These are responsible for remeasuring the work as executed and for preparing the information for interim payments. They will also be involved in preparing the final account and contractual claims. This function is to ensure that the contractor gets paid the correct amount at the correct time.

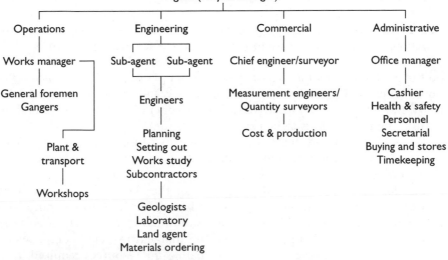

Figure 11.1 Contractor's site organisation

Operations management

This aspect of the site management is responsible for making sure that the temporary and permanent construction works are carried out in accordance with their engineer's instructions. On some projects a general foreman will supervise a number of trade foremen to supervise the supply and use of materials, the activities of the workmen and the allocation of mechanical plant and equipment. On the larger projects works managers are employed for this purpose, often at a similar status to that of the sub-agent.

Administrative management

Site administration will vary in size from a single administrator to larger civil engineering construction projects where it is necessary to employ an office manager, who oversees several different functions, these will include the secretarial work, cashiers and stores managers, site welfare, health and safety and labour relations.

Planning and programming techniques

A programme or schedule is developed by breaking down the work involved in a construction project into a series of operations which are then shown in an ordered stage-by-stage representation. Without a programme of work which specifies the time and resources allocation for each stage of the project, the execution of the contract will be haphazard and disordered. There are several methods used in programming, which may be broadly classified as follows:

Bar or Gantt charts

This method of planning and programming is used extensively in the construction industry and is easy to comprehend. Each site operation is allocated a timescale. The charts generally show the operations in the first column on the left-hand side of the chart. The contract period, from start to completion, including the defects correction period, is shown along the top of the chart. Bar lines are then inserted against the operation and under the time period when the work is to be carried out on site.

The start, finish and duration of each activity can be clearly seen, as can the way in which the different activities relate to each other. However, bar charts do not show the dependency of one activity on another.

The linked bar chart is a development of the above, showing the links between an activity and the preceding activities that have to be completed before it can start. Similarly, the links are shown between the activity and the succeeding activities that are dependent on that activity being completed.

The bar chart is a useful tool for calculating the resources required for a project. The resources for each activity can be calculated and totalled to provide a resource aggregation for the activities and the project as a whole.

Network analysis

Network analysis is an operational research technique that has been used extensively on civil engineering projects. It is often referred to as the critical path method (CPM) or programme evaluation and review technique (PERT). Whilst the principles involved in the two techniques are similar, there are some differences in their application. PERT was originally designed to cope with uncertainty by considering three estimates – optimistic, most likely and pessimistic – for each activity. Network analysis imposes a more rigid discipline during the development of an overall plan for a project, since it is necessary to list a complete set of site operations. Also, the relationship between preceding and following operations must be precisely decided.

The linkage between preceding and succeeding activities, combined with a set of arrows to represent the bars of the bar chart, gave rise to a simple network diagram. This formed the basis of a network analysis, i.e. the critical path method, which identifies the longest irreducible sequence of events and also lends itself to manipulation because the data is held in computer files. This results in a powerful planning technique which can be quickly updated. It also defines quickly those parts of the programme which could benefit from the use of increased resources and thereby benefit the project. As networks are rarely the best method for communication, the output of the analysis is often presented as a bar chart. Network analysis has been used for many large and complex projects. It has been claimed that time savings in excess of 40 per cent have been achieved by its use.

Line of balance

This is a planning technique that is useful for repetitive work. The basis of the technique is to find the required resources for each stage or operation, so that the following stages are not interfered with and the target output can be achieved. The technique has been used for mass house-building and, to a more limited extent, for roadworks.

Resources

The time taken to complete an activity in the programme is dependent on the resources allocated to that activity. The approaches used in assessing the required resources can be based on completing the project in a given time or completing the project with specified limited resources. Once the level of resourcing has been finalised, the resource demands are smoothed if necessary by rescheduling activities to ensure an acceptable overall demand for the project.

There is a measure of uncertainty in estimating the time for each activity, particularly as late delivery of materials and adverse weather can delay work in progress. Probabilistic distributions have been used for the generation of the most likely times for activities; an approach that was used in early projects was the project evaluation and review technique (PERT). Generally, there should not be difficulty in ensuring that the materials, properly ordered in good time, are delivered to site in time for the required activity to be undertaken. Some materials have long lead times and these may prove expensive if early delivery is needed; however, an accommodation can sometimes be found by agreeing a substitute material. A less conscientious contractor may decide to ignore the difficulties and seek to obtain a claim at the end of the contract. The position with regard to the flexibility of the workforce has changed in recent years. When contractors operated a number of sites in an area with their own workforce forming the bulk of the manpower, there were ample opportunities for operating flexibly. However, the arrival of a greater measure of subcontracting has meant that workforces are less flexible than a directly employed workforce. There is less flexibility in day-to-day site management and adaptability has been replaced by a greater measure of pre-site planning.

The labour-intensive nature of contracting has brought about a state of affairs where the use of plant and machinery in the construction industry has traditionally been less than in manufacturing. Nonetheless, techniques do require the use of plant in which firms have directly invested or obtained by means of plant-hire. The equipment used is necessarily mobile and downtime or idle-time is minimised to ensure that the return on the investment is achieved. Nonetheless, the amount of time that plant is unused can be high; as much as 90 per cent. This places a considerable capital burden on the contractor. The plant-hire industry ensures the more intensive use of plant and the hirer also supplies skilled operators. Plant-hire is responsible for supplying over half the needs of the construction industry.

Monitoring and control

With a programme of work and the resource requirements for each activity having been determined, it is possible to monitor the construction work as it progresses. In practice, updating will take place and control will be exercised. This will entail the rescheduling of activities and the revision of resources. The emphasis on greater control, particularly cost control and site activities, has been one of the more important developments in the construction industry during the past decade or so. It was brought about by the decline in workload in the early 1970s, which caused contracting firms to focus more closely on their profitability, thus paying greater attention to site management and control.

The planning model is often used to explore the overall development of the project before work on site is undertaken. This assists in investigating the influence of different construction techniques and the timing of the individual activities to optimise the use of resources.

Ensuring that the works are constructed to the specified level of quality is essential. This extends from the initial setting out of the project to the inspection, storage, handling and incorporation of the specified materials. All site operations should be governed by appropriate safety measures, which start with a safe design and erection procedure for permanent and temporary works. Safety control is a matter of comparing site statistics with comparable statistics in addition to locating areas where it is necessary to take action. This is not at all easy to do. Construction is the largest single contributor to fatal accidents under the Factories Act.

Site layout and organisation

Each construction site is unique. It will have its own defined shape, topography and geological strata. It may have access to engineering services such as water, electricity, gas and sewage or these may be located some distance away and need to be either temporarily or permanently installed. On some sites these utility supplies may transverse the site, resulting in difficult site operations.

A site layout plan should be prepared, showing the best locations for the various facilities. On large or extended sites these may need to be duplicated for the sake of efficiency. The following factors are important.

Access

The requirements for this are usually specified in the contract documents (specification). Generally, access to the site will only be allowed at certain points. A motorway project that might transverse a number of minor roads is unlikely to be accessed from each of these roads. On large projects, road signs are erected that direct the contractor's traffic to certain site entrances. This helps to reduce congestion and inconvenience to third parties in the area where the site is located. Often the access and egress to a large site will be via a one-way traffic system.

The temporary access to a site is usually constructed of hardcore or railway sleepers. Wherever possible, when this is practicable and allowed, a temporary access road can be used as the foundation for a more permanent road, usually forming the road base construction.

On large earthmoving contracts, where large amounts of spoil need to be taken off-site to a tip, the contractor will need to seek permission to use public roads. The shortest route might not always be possible due to traffic restrictions. In some cases it may be necessary to construct temporary roads and Bailey bridges. Where they are to be used over long periods they may also require some form of drainage to be installed. All routes will need to be carefully maintained and, in the case of public highways, the surfaces must be kept clean.

Materials storage and handling

The principal objective is to minimise wastage that might arise through:

- careless use
- multiple handling
- poor storage
- theft
- vandalism

Secure stores and compounds must be provided for tools, equipment, plant spares, goods, components and materials. Racks should be provided for materials such as scaffolding and stillages built for oil drums. Materials that are susceptible to damage from the weather, such as cements, should be stored in dry weatherproof buildings. Sands and aggregates should be stored on hard surfaces in order to reduce their wastage and contamination.

Where a structure is being constructed with compartments, it may be possible to use the newly completed rooms for the storage of materials that are easily damaged or stolen.

The security of construction sites can be a problem, particularly in the case of elongated sites where protection is difficult and unauthorised access easy. The use of locked buildings, substantial fences with gates and the employment of security firms with guard dogs, night watchmen or a visiting patrolman will need to be considered.

The methods used for materials handling should be efficient and effective and a greater use therefore needs to be made of mechanical plant. The methods used should reflect the need to protect the goods, materials and components while in transit around the site.

The wastage of goods and materials is still high – often higher than estimators and site managers are prepared to admit. Studies on wastage have been completed by many different organisations. Although these have frequently reported high wastage levels, the actual figures are often much higher. Misuse of materials is cited as one of the main causes.

Site accommodation

The siting of the site office accommodation requires a careful compromise between the need to overlook the site operations and freedom from the noise and dust generated by construction work. The size of the temporary offices will be influenced by the number of staff and the activities that they need to perform. There some privacy will be necessary for discussion, as will accommodation for site meetings.

The specification will usually detail the requirements of the resident engineer. The contractor will also have to decide on the layout of space needed for the agent and the agent's staff.

Temporary site buildings now require little more erection or dismantling than a caravan. Timber sectional offices are almost a thing of the past. Car parking will also have to be considered. On a congested site this may be difficult.

Temporary mess rooms for the construction operatives, canteens, drying rooms and toilets will also be referred to in the specification. Such welfare facilities have now become an important part of construction sites.

Plant and workshops

The choice of mechanical plant represents a major decision on most construction sites. The contractor's method statement will identify the way in which it is proposed to carry out the construction work. Many site operations now rely upon a good balance between the use of mechanical plant and people. The various options of hire, purchase or leasing will need to be considered, compared and evaluated.

The workshops for carpenters and machine fitters must be selected on the basis of a balance between easy access and closeness to the construction operations and the importance of alleviating congestion on the site. Where possible, workshops should be located where engineering services can be easily accessed. The installation of temporary services will need to be agreed with the various authorities. The availability of water pressure or electricity voltage may influence the amount and type of workshop plant that can be installed.

Special problems

The contractor will need to consider the peculiarities of each individual site. These include:

- Confined sites where access may be difficult require the detailed planning of material deliveries, offices at a high level and careful planning of the site operations.
- Elongated sites may need a number of office and workshop compounds and special attention will have to be given to matters of security.
- Tall structures require a carefully sited tower crane, the installation of passenger lifts and goods hoists and a more meticulous attention to safety.
- Staged completions involve the handing-over to the employer of the completed stages of a project. This will influence the planning of the whole project and the possible relocation of some of the temporary works.
- Adjoining owners can cause difficulty for the construction operations. This is particularly so in respect of demolition or pile driving. Special care needs to be exercised to avoid possible claims for damage or trespass.

Incentive schemes and motivation

Bonus payments have been used in the construction industry for many decades. They have been the cause of much conflict between operatives and management but are believed to result in a project being completed sooner and at a lower overall cost to the contractor. Benefits are also thought to arise from the resulting improved site organisation and management as operatives who are earning bonuses will not tolerate working practices and conditions that cause delay to their work.

A good incentive scheme should be one that is easily understood by those concerned. It should provide operatives with reasonable opportunities to increase their earnings above the basic wage. The formulation of such a scheme requires both technical and psychological skills. Whether or not the operatives benefit significantly from the bonus system is arguable. The construction trades unions believe that incentive schemes tend to depress the basic rate of payment and it is said that some would welcome a higher basic rate linked to the discontinuance of incentive schemes.

Despite the fact that there are a number of motivating factors for people in employment, construction management pays particular attention to wage levels and less attention to the conditions under which the operatives are employed. In the case of salaried staff company cars, pension schemes and other benefits are often offered as forms of incentive, although few of these are linked directly to increased productivity.

Health and safety

There has been concern for many years over accidents and working conditions in construction work. A Public Enquiry was held in 1904 and the first Building Regulations were passed by Parliament in 1926. These regulations were revised and updated in 1931. The 1937 Factories Act empowered the Minister of Labour to make regulations for the construction industry, but it was not until 1948 that the Building (Safety, Health and Welfare) Regulations were brought into force. These regulations did not apply to civil engineering works. During the post-war years the construction industry expanded and the accident rate gave cause for concern. In March 1962, the Construction (General Provision) Regulations and the Construction (Lifting Operations) Regulations replaced parts of the earlier regulations. These were extended to civil engineering as well as building. Two further sets of regulations, the Construction (Health and Welfare) Regulations 1966 and the Construction (Working Places) Regulations 1966 were brought into operation. In 1974, a major piece of legislation, the Health and Safety at Work Act 1974 was introduced. Further regulations have since been added. The Health and Safety at Work Act covers all people at work with few exceptions, whether employers, employees or the self-employed. It also protects the general public in cases where the work activities of others may affect its health and safety. More recently the Construction (Design and Management) Regulations 1994 have been introduced and these are now a formal part of the ICE Conditions of Contract (see Chapter 15).

The risk of a fatal injury in construction is about 4 times greater than in manufacturing industry; the risk of a major injury is 1.5 times greater in construction. Of the fatal accidents, 74 per cent occurred in the building sector of the construction industry, i.e. construction, demolition and maintenance of houses, commercial and industrial premises. Approximately a quarter were associated with the civil engineering sector, i.e. construction and maintenance of roads, sewer pipelines, sea and harbour defences and large petrochemical, oil and gas installations. The level of fatal accidents is not falling at the same rate in construction as in manufacturing.

Because of the nature of the construction industry, there are particular hazards for those who work in it. The working environment on a construction site is subject to constant change and, as the working activities change, new risks are involved. The casual and short-term nature of employment compounds these risks. In contrast, most manufacturing is undertaken in a more controlled environment, with specified activities undertaken by a relatively unchanging workforce. Were the nature of construction to change, with a greater bulk of component parts fabricated under more closely controlled conditions and brought to site for erection only, safety benefits could result. Increasing industrialisation also has benefits for quality and productivity.

Further contributory factors to the higher incidence of accidents in construction could be the multiplicity of small firms and the widespread use of subcontractors.

CHAPTER 12

SITE COMMUNICATIONS

Introduction

The construction of engineering structures and similar projects utilises a range of different resources such as people, machines, money, materials, mechanical plant, management and methods. It seeks to combine these together in the most effective, efficient and economic manner.

The construction industry is unique. It was one of the few industries to separate design from production, although this is changing with the increase in design and construct projects. It also undertakes the bulk of its work on the employer's premises, i.e. the construction site. The industry does not build prototypes and, whilst projects may appear to be similar, they are different in many respects.

Site management organises, informs, co-ordinates, orders, instructs and motivates others to undertake these tasks. The effectiveness of performance will depend upon the ability to listen, read, speak and write. More importantly, effectiveness will be evident in being understood by others and in understanding another person's point of view.

Site communications

Site communications therefore includes gathering information to ascertain the needs of those involved with the project. This is a necessary tool of management to integrate the functions of departments and a vital link between manager and subordinate in order to get the job completed.

Communication principles

Communication is a key to success in the work of a site agent. Communication skills and strategies are necessary to share ideas and experiences, to find out about things and to explain to others what is required. Developing ways of communicating, including language and body signals, is essential in expressing feelings and insights. Learning to communicate effectively may mean the difference between barely coping with life and actively shaping the world. Some of the components of good communications are listed below:

- knowing what to say
- gaining the attention of the listener
- establishing and maintaining relationships
- knowing the listener's likes and interests
- choosing how to communicate from a range of options
- skill in communicating
- choosing when and where to communicate
- being clear, brief and coherent
- being an active listener
- understanding and clarifying messages received
- not being easily distracted
- knowing how to close conversation or communication

Effective communication starts with a purpose or objective to achieve in transferring the information. This should be done in the most suitable form to suit the receiver. It also needs to be the most effective way. Provision should also be allowed for feedback and the need for action on that feedback appreciated. The points that should be present in all communications should be:

- clarity, i.e. easily understood
- presentable, i.e. creating a favourable impression
- new information, i.e. to attract interest
- drive, i.e. demanding the necessary action
- tone, i.e. creating a responsive attitude
- feedback, i.e. ensuring that the transfer of information is achieved

Forms of communication

There are three basic methods of communication.

Written

This takes the form of letters, reports, bills of quantities, specifications, site instructions, British Standards. This method is used when the subject matter is complex, important or is likely to have possible legal implications and where a permanent record is required for future reference.

Visual

This can take the form of films, slides, posters, graphs, charts. It includes the project drawings and works programmes. This sort of communication often has a greater impact upon the receiver. Where messages are less complex, this form of communication is the most effective.

Oral

This method is instant and often generates an immediate response. Attitudes and behaviour can be also observed. The face-to-face confrontation is appropriate when the subject matter may be difficult or disagreeable. It is also used in those circumstances for simple, less important and informal messages.

Barriers to communication

An organisation that is suffering from poor morale or lacks confidence in its future is likely to see problems in communications increase. In these circumstances there is sometimes a lack of will to communicate effectively. However, nearly all barriers to good communication can be classified as follows:

Physical

These might include disability on the part of the receiver, such as deafness or blindness. It might include noise on site, poor telephone lines and layers of site management. These can sometimes have the effect of distorting the information that is being transmitted.

Psychological

These are the biggest cause of communication breakdown. They affect the attitudes, feelings and emotions of the receiver. Sometimes, individuals will only hear what they expect to hear. Preconceived ideas are used to interpret the information, in ways not anticipated by the giver of the information. Feelings and emotions affect the ability to receive a true message. When worried, the receiver feels threatened. When he or she is angry, the information may simply be rejected.

Intellectual

These affect concepts and perceptions that may have been built up over a lifetime of work. Difficulties may arise because the information was not transmitted properly in the first place. It may use excessive technical jargon or attempt to use words that are out of context.

Site information

In addition to the everyday running and organising of a contract, it is also the site agent's responsibility to maintain accurate records of the important happenings on site. This information should be properly recorded, so that it can be quickly retrieved for future use. The types of site records normally kept include:

- daily reports
- site diaries
- materials received sheets
- advice of variations
- daily labour allocation sheets
- drawing registers
- confirmation of verbal instructions
- weather reports
- daywork records
- subcontractor files
- site correspondence

If this information is to be of value for future reference it must be:

- Precise: it should be clear, straightforward, as simple as possible and well thought out
- Accurate: true and correct in all of its details
- Definite: future users should be left in no doubt as to what the message means
- Relevant: to the particular situation and the people who will use it
- Referenced: wherever possible the information should be linked to other documents being used by the contractor

Site meetings

Site meetings are held for a variety of different purposes. They may be held with the contractor's staff only, to establish how the works should carried out or to deal with an internal matter that has arisen in connection with the project. They often involve other parties such as subcontractors and suppliers. The general site meeting would generally include the employer, engineer, quantity surveyor, other consultants, subcontractors and the main contractor. These meetings frequently take place on a monthly basis, depending upon the size of the project and the issues that may need to be resolved. The site meeting is an opportunity to pass information from one party to another or to discuss how to tackle a problem that has arisen. The normal site meeting is also used for decision-making purposes and it may be used to try to persuade the parties involved of a different way of solving a particular problem. In essence it covers the questions:

- 'What do we need to know?'
- 'How are we getting on?'
- 'What's wrong and how can this be corrected?'

Aspects of meetings preparation are listed below:

- Determine meeting's objectives.
- Draw up attendance list.
- Decide on the time and place.
- Determine the style of the meeting.
- Circulate agenda and other information.
- Identify whether special facilities are required (e.g. audio/visual).
- Rank and set times for each item.
- Assess possible areas of conflict.
- Discuss items with individuals in advance where necessary.
- Determine sort of minutes that are required.
- Adequately brief participants.
- Other matters, refreshments, car parking, etc.

Frequently, the engineer will chair and direct the site meeting and prepare an agenda. This is usually done in consultation with the contractor. The agenda will follow the format of a business meeting, with apologies for absence, matters arising from the previous minutes, any other business and the date and place of the next meeting. Items of a routine nature, such as the main contractor's report and special items that address current issues will also be included.

The minutes of the meeting are the follow-up and, as such, are vital to the effectiveness of the meeting. The minutes express the collective views of the meeting. They should also identify the areas where responsibility for further action lies. They may include an action plan in order to minimise future problems or difficulties that might be encountered. The list below gives points to remember that should help meetings to flow more easily and become more productive:

- Prepare an agenda and establish rules of procedure.
- Start promptly and be brief.
- Listen actively; practise this skill, it leads to good questions, improves group understanding and keeps meetings good humoured.
- Give direct replies; keep the meeting to the point to avoid wasting time.
- Clarify issues: 'Are you suggesting that you provide this material by the end of the month?'
- Summarise progress.
- Restate important points.
- Be prepared to change strategy if necessary.
- Be supportive: 'That sounds like good idea'.
- Confront issues: 'Are we really prepared to . . .'
- Question critically: 'What exactly do you mean?'
- Provide accurate supporting information.
- Do not allow the meeting to be interrupted by telephone calls, etc.
- Avoid interrupting.

- Do not be afraid to make your feelings known.
- Refrain from distracting behaviour.
- Do not talk to others during presentations.
- Never lose your temper, except deliberately.
- Do not embarrass others.

Site diaries

The site diary, if well maintained with the correct sort of information, is a very useful document, retained by the contractor. It will contain information that can be used where disputes occur between the different parties concerned with the project. Site diaries have been used in litigation as valuable evidence to help substantiate a case put forward by the contractors. The diary should record information that generally does not warrant separate records being kept. This will vary with the size and type of project, the nature of the employer and the information records kept by the contractor. Figure 12.1 provides a typical form that can be used for this purpose.

- Weather conditions: These will have a major influence upon the progress of the works. They must therefore be recorded, especially where they cause a delay or suspension of the works.
- Drawings received: The site agent must ensure that there is an adequate system for the receipt and recording of architect's and engineer's drawings. Although there will be a drawing register, the site diary should also record when this information is received. There will also be times when information will be requested from the consultants. This should always be in writing, giving sufficient notice to avoid the possibility of delays. The late receipt of information is one of the reasons available for an extension of time. In addition, if the delay in the receipt of such information causes the works to be suspended for longer than the period named in the contract, then the contractor may determine the contract. The contractor in each case must have requested further instructions and drawings and the date of that request will be the start of the period of delay.
- Engineer's instructions: Instructions will be given by the engineer, both orally and in writing. These need to be recorded in the site diary under the heading of site instructions. It should be remembered that, in order to be valid, all instructions must eventually be in writing. The site agent therefore needs to record precisely when the instruction was given.
- Variations: These are a special type of engineer's instruction since they may alter or modify the design, quality or quantity works shown on the drawings and in the specification and bills of quantities. A variation may also alter the conditions under which the work is to be carried out.
- Dayworks: The site agent should record in the site diary reasons why some of the work should be paid for on this basis. Where dayworks are envisaged, the engineer should be informed in order that the labour and materials

DETAILS: DATE:

_____

1. Weather conditions   Temperature

_____

2. Drawings received   Number

_____

3. Instructions: Oral/Written   Given by

_____

4. Variations received   Number

_____

5. Dayworks: Reasons/Descriptions   Sheet numbers

_____

6. Delays: Reasons   Labour on-site required

_____

7. Lost time: Reasons

_____

8. Urgent requirements

_____   Visitors to site

9. Unusual occurrences

_____

Figure 12.1 Site diary
Source: adapted from W. H. Davies, *Construction Site Studies*, Butterworth, 1982

used can be properly checked. The inclusion of large numbers of dayworks in the site agent's diary may lead towards the conclusion that the nature of the project has changed from that which was originally tendered.

- Delays: The contractor will have prepared a programme for the construction of the project. The programme will indicate how the contractor intends to carry out and complete the works. It also makes the engineer aware of the dates when the contractor requires the various pieces of information. Measuring progress against the programme provides the contractor with the best indication that the works are being constructed as planned. Where the progress shown is different, the reasons will need to be established.
- Lost time: A record of any delays will be made in the site diary at the time that they actually occur. This will be the most accurate record of these events. Where the delays are due to factors that the contractor should have controlled, these will need to be remedied at the contractor's own expense. This will then avoid liquidated damages being applied for late completion.
- Urgent requirements: The site diary will also be used to remind the site agent of any urgent requirements. These may include a note in advance, of items that are expected for delivery or action on a certain date.
- Unusual occurrences: Any unusual happening on the site will need to be recorded for possible future reference by the contractor. Disputes may occur between the parties concerned and it is important to establish as soon as possible why such disputes have arisen. The following are typical examples:

  - Local stoppage on site for two hours due to a disagreement over bonus payments.
  - Mr R Taylor fractured his ankle. An accident report has been completed.
  - Intruders entered the site last evening. No damage or theft occurred. Police have been informed

- Labour records: The site agent will need to keep a careful record of both planned labour requirements and that actually employed on site.
- Visitors to the site: Visitors should always be recorded, especially those who might have a direct impact upon the work. The contractor cannot refuse reasonable entry to the site for the engineer or engineer's representatives. The civil engineering employer owns the site and is therefore allowed access. This must be done to cause as little inconvenience to the contractor as possible. Statutory inspectors such as those concerned with aspects of health and safety and similar officials have a legal entry to the site at any reasonable time.

Recording of information

If the site dairy is to be of any value then it is important that the events are entered each day in a logical, careful and legible manner. The information will thus provide an accurate assessment of the site's daily progress together with the labour and plant that has been used. Any matters that affect the following must be regularly and accurately recorded:

* completion date
* costs
* quality and standards of work
* contractor's performance

The site diary must be completed daily so that the salient points are not permanently forgotten. The completion of the task will require some self-discipline on the part of the site agent, particularly when other aspects of work are requiring attention. The information should be accurate and represent a fair picture of the day's events. Exaggerated information or hearsay remarks should not be included. Where proper and adequate records are not maintained, the contractor may suffer because of:

* loss of reputation of a well-run organisation
* imposition of liquidated damages through late completion of the works
* refusal of additional payments for losses incurred
* unfair termination of the contract due to employer dissatisfaction
* levy of further damages to redress the employer's loss

PART FOUR

ICE CONDITIONS OF CONTRACT

CHAPTER 13

GENERAL MATTERS

Introduction

This chapter includes:

- engineer and engineer's representative (clause 2)
- property in materials and contractor's equipment (clause 53)
- vesting of goods and materials not on site (clause 54)
- urgent repairs (clause 62)
- determination of contractor's employment (clause 63)
- war clause (clause 65)
- application to Scotland and Northern Ireland (clause 67)
- notices (clause 68)
- tax matters (clause 69)
- value added tax (clause 70)
- special conditions (clause 72)

Engineer and engineer's representative (clause 2)

Duties and responsibilities of the engineer

The engineer must carry out the duties that are:

- specified in the contract
- implied from the contract

The engineer is deemed to have the authority from the employer to carry out the duties and responsibilities of the contract. However, the engineer does not have any authority to:

- amend the terms and conditions of the contract
- relieve the contractor from any obligations under the contract

Named individual

The engineer is defined in clause 1 (1)(c) and is referred to in clause 2 (2). This means the person, firm or company appointed by the employer to act as engineer for the purposes of the contract. The engineer is named in the Appendix to the form of tender. This may not be a single named individual. If this is the case, the firm or company appointed must write to the contractor prior to the works commencement date to nominate an individual chartered engineer who will act in the capacity of the engineer. Where this individual engineer is subsequently replaced, the contractor must be informed of the name of the replacement engineer.

Engineer's representative

The engineer's representative is responsible to the engineer. The representative has no authority to:

- relieve the contractor of duties and obligations
- order work involving delay or extra payment (unless expressly provided)
- make any variation of or in the works (unless expressly provided)

Delegation by engineer

The engineer can at any time delegate to the engineer's representative duties and responsibilities that are vested in the engineer. These can also be revoked at any time. Any such delegation:

- must be in writing and delivered to the contractor (Note that clause 1 (6) allows for many different mediums of writing.)
- will remain in force until the revocation is notified in writing to the contractor
- must not be given in respect of certificates issued under:
  Clause 12 (6) Adverse physical conditions
  Clause 44   Extension of time
  Clause 46 (3) Accelerated completion
  Clause 48   Certificate of substantial completion
  Clause 60 (4) Final account
  Clause 61   Defects correction certificate
  Clause 63   Determination of the contractor's employment
  Clause 66   Settlement of disputes

Assistants

The engineer or engineer's representative may appoint any number of assistants to help them carry out their duties. The contractor must be informed of:

- their names
- duties
- scope and authority

These assistants will not generally have authority to issue instructions to the contractor. Where these are required to enable them to carry out their duties, such as the acceptance of materials or workmanship and are in writing, they will be deemed to have been given by the engineer's representative. Where the contractor is dissatisfied with such instructions, the matter will be resolved by the engineer's representative.

Instructions

Please refer to Chapter 17.

Reference to dissatisfaction

Where the contractor is dissatisfied with an instruction from the engineer's representative, the contractor is entitled to refer the matter to the engineer for a decision.

Impartiality

One of the engineer's roles is to act impartially between the two parties to the contract, i.e. the employer and the contractor. The engineer is employed on behalf of the employer and receives fees from the employer. There will be a separate contract between the employer and the engineer. The extent of impartiality will vary. The point of this clause is to enable the engineer to act in a professional capacity, supporting the contractor where the contractor has a good case.

Property in materials and contractor's equipment (clause 53)

Vesting of contractor's equipment

All of the contractor's temporary works and materials used for temporary works are deemed to be the property of the employer while on site. They should not be removed without the written consent of the engineer. This consent should not unreasonably be withheld, particularly where they are no longer required for the completion of the works.

Liability for loss or damage to contractor's equipment

The employer is not liable for any loss that might be sustained to the contractor's temporary works, goods or materials other than that outlined in clause 22 (Damage to persons and property) and clause 65 (War damage).

Disposal of contractor's equipment

The contractor must remove all equipment, temporary works, goods and materials as required under clause 33 (Clearance of site on completion). This must be done within a reasonable time after the completion of the works. Where the contractor fails to remove any of these items, then the engineer may allow the employer to dispose of them. The employer is allowed to deduct costs and expenses from the proceeds of any sale before paying the balance to the contractor.

Vesting of goods and materials not on site (clause 54)

Where the contractor is able to secure payment for goods or materials off-site, the transfer of their ownership must be passed from the contractor to the employer. This provision covers only those items that have been listed in the Appendix to the form of tender. The clause is therefore intended to cover:

- only those materials that have been specifically manufactured for the project and are substantially ready to be incorporated within the works. The costs of manufacture may be considerable, and fabrication may take place some considerable period of time prior to their incorporation within the works
- goods and materials that are the property of the contractor
- property that can pass unconditionally to the contractor

Action by contractor

In order for the transfer of property to take place, the contractor or the supplier must take the following action:

- provide documentary evidence that the property in the goods or materials is vested in the contractor
- suitably mark or identify the goods and materials to the effect that their destination is the named site and that they are the property of the employer, unless they are stored on the contractor's premises
- set aside the marked goods and materials to the satisfaction of the engineer
- send the engineer a schedule listing the goods and materials and their respective values and inviting the engineer to inspect them

Vesting in employer

The engineer must then approve the transfer of ownership of any goods and materials described above, to become the absolute property of the employer. Their possession by the contractor is for the sole purpose of delivering them to the site and incorporating them into the works. The vesting does not:

- prohibit the engineer from rejecting goods or materials that are not in accordance with the provisions of the contract
- remove the contractor's responsibility (for which additional insurance may be required) for any loss or damage to the goods through storing, handling or transportation

Lien on goods and materials

The contractor, subcontractor or other persons do not have any lien (a hold by one person over another's property) on any goods or materials which have been vested in the employer. The contractor must ensure that the title of the employer is brought to the notice of any other interested party.

Delivery to the employer of vested goods and materials

If the contractor's employment is determined under clause 63, then the contractor must deliver to the employer those goods or materials which have been vested in the employer. Where the contractor fails to do so, the employer may enter any of the premises owned by the contractor or subcontractor and remove the goods or materials. The costs of this action is also to be recovered from the contractor.

Incorporation in subcontracts

The provisions of clause 54 must be incorporated in all subcontract provisions arranged by the contractor.

REMEDIES AND POWERS

The civil engineer has wide-ranging remedies and powers under the following conditions:

Urgent repairs (clause 62)

Where accidents or failures of some part of the works occur, the engineer can issue instructions for their repair. This may occur during the execution of the works or during the defects correction period. The engineer must inform the contractor in writing as soon as possible after the occurrence of such an event.

Where the contractor is unable or unwilling to carry out this work, the employer may employ other firms to do the work. Where the responsibility for the repair is at the contractor's expense, the additional costs or charges properly incurred can be deducted from monies due to the contractor.

Determination of contractor's employment (clause 63)

Where a contractor defaults in one of the following ways:

- becomes bankrupt or has a receiving order or administration order made against him (other than voluntary liquidation for the purpose of amalgamation or reconstruction)

- assigns the contract without consent in writing from the employer
- has an execution levied on goods which is not stayed or discharged within 28 days
- has abandoned the contract
- fails, without a reasonable cause, to commence the works in accordance with clause 41
- Suspends the progress of the works without a reasonable cause
- fails to remove from the site condemned goods or materials or defective work after receiving instructions from the engineer to do so
- Fails to proceed diligently with the works despite previous written warnings from the engineer
- is persistently in breach of obligations under the contract

then the employer may, after giving 7 days' notice in writing to the contractor:

- specify the default
- enter the site and works
- expel the contractor

This will not necessarily relieve the contractor from the obligations or liabilities of the contract. Where appropriate, the employer should allow the contractor the opportunity to remedy the default.

Completing the works

In the case of the determination of the contractor's employment, the employer or a new contractor may, in order to complete the works, use the contractor's:

- equipment
- temporary works
- goods and materials

These are deemed to have become the property of the employer under clause 53 and clause 54. The employer may at any time sell these items and apply the proceeds of sale towards any sums that may become due from the contractor under the terms of the contract.

Assignment to employer

Within 7 days of determination, the contractor must assign to the employer any benefits regarding the agreement to supply goods and materials that the contractor may have arranged.

Payment after determination

Any payments due to the contractor will not be made until the end of the defects correction period. The amounts due are subject to adjustment for the expenses and costs associated with employing another contractor to complete the works. The contractor is only entitled to receive what would normally have been due for completing the works up to the time of determination. In practice, under these circumstances, the costs of completing the works are often far in excess of any amounts outstanding to the contractor, unless there are large sums involved in respect of the contractor's equipment. It is therefore more common for this debt involved to be recovered by the employer from the contractor.

Valuation at date of determination

As soon as practicable after determination, the engineer must calculate:

- the amount reasonably due to the contractor in respect of the works so far completed
- the value of unused goods and materials, equipment and temporary works

War clause (clause 65)

Works to continue for 28 days at outbreak of war

This clause is concerned with the outbreak of war involving Great Britain. War does not need to be declared for this clause to become effective; it comes into effect on the announcement of a general mobilisation of the armed forces. The contractor should, as far as is physically possible, continue to execute the works in accordance with the contract for a period of 28 days.

Effect of completion within 28 days

If the works are completed within the 28 days or sufficiently so to allow the project to be usable, then the contract will be deemed to have been completed and

- the contractor will, in lieu of fulfilling the obligations under clause 49 and clause 50 (Outstanding works and defects), be paid a reduced contract price
- the employer will not be allowed to withhold payment under clause 60 (6)(c) of the second half of the retention money

Right of employer to determine contract

Where the works are not completed, the employer is entitled to determine the contract, other than in respect of clause 66 (Settlement of disputes) and clause 68 (Notices). The contractor can be given notice after the expiry of the 28-day period above.

Removal of contractor's equipment on determination

If the contract is determined, the contractor must remove all the contractor's equipment from the site as soon as possible and allow subcontractors to do the same. Where the contractor fails to do this, the employer has the same powers as in clause 53 to dispose of the equipment.

Payment on determination

Where the contract is determined under this clause, the contractor will be paid the following by the employer:

* the amounts of any preliminary items that have been completed
* the costs of materials and goods reasonably ordered for the works, where the contractor is legally liable to accept their delivery
* expenditure that has reasonably been incurred by the contractor in the expectation of completing the works
* any additional sum payable under sub-clause (6) described below
* any reasonable cost of removal under sub-clause (4) (Removal of contractor's equipment on determination)

Provisions to apply from outbreak of war

Whether the contract has been determined, due to an outbreak of war or not, the following provisions will apply:

* The contractor is under no liability for damage to the works or to property belonging to the employer. (The contractor remains responsible for his own property or that on hire.)
* The employer must indemnify the contractor against all such liabilities and claims.
* Where the works are destroyed due to an outbreak of war, the contractor is nevertheless still entitled to payment for the work that has been satisfactorily completed.
* The contractor may be requested to repair the damaged works, as required by the engineer. This will be paid for on a cost plus basis.
* Where the contract includes a contract price fluctuations clause, special arrangements are made to deal with any increases or decreases in costs.
* If the costs of the works are increased or decreased because of any statute or statutory instrument, the contract price will be adjusted accordingly.

Application to Scotland and Northern Ireland (clause 67)

If the works are located in Scotland, the contract must be interpreted in accordance with Scottish law, which has some significant differences from English law. Similar interpretation is provided for projects that are to be constructed in Northern Ireland. It should be noted that English law generally applies to Wales.

Notices (clause 68)

Notices served by the employer on the contractor within the terms of the contract must be in writing. They should be sent to the contractor's principal place of business or its registered office. Notices served by the contractor on the employer are to be dealt with in a similar manner.

Tax matters (clause 69)

Labour tax fluctuations

The rates and prices contained in the bill of quantities are deemed to be those current at the date for return of tenders. These cover taxes, levies, contributions, payments and refunds. They do not therefore take into account any changes that might be known but have not yet been implemented. If, after the date for return of tenders, changes are made to these items, the contract price can be adjusted accordingly. The contractor must supply the information necessary to support any increase or decrease to the contract price.

Value Added Tax (clause 70)

The contract price is deemed to be exclusive of VAT. All certificates issued by the engineer are therefore exclusive of VAT.

Engineer's certificates net of Value Added Tax

In addition to the payments due under certificates, the employer must also separately identify and pay the contractor any Value Added Tax that is properly chargeable by the Commissioners of Customs and Excise.

Disputes

Where disputes arise in respect of VAT, the contractor and employer should offer each other support and assistance in resolving the dispute, difference or question.

Clause 66 not applicable

Clause 66 (Settlement of disputes) does not apply to any dispute, difference or question arising under this clause. Any disputes regarding VAT are dealt with by the Commissioners, whose decision is final and not subject to arbitration.

Special conditions (clause 72)

Special conditions, such as the price fluctuation clauses which are to be incorporated within the contract are to be added after clause 72. Only the standard additional clauses can be added and not ones that may have been specially devised by the employer. Attempting to add other clauses or amending or removing clauses from the Conditions of Contract can have the effect of making the contract null and void.

CHAPTER 14

CONTRACTOR'S OBLIGATIONS

Introduction

This section represents the largest number of clauses in the ICE Conditions of Contract (clauses 8–35). There is no clause 34. Some of the clauses have been considered elsewhere. Clause 14 (Programme) has been dealt with in Chapter 5 (Contract documents), and clause 15 (Contractor's superintendence) in Chapter 10 (Parties and process).

Contractor's general responsibilities (clause 8)

The contractor's responsibilities under the contract are to:

- construct and complete the works
- provide everything necessary, i.e.
  labour
  materials
  contractor's equipment
  temporary works
  transport on and off the site

The permanent works are defined in clause 1 (n) of the conditions of contract:

Permanent works means the permanent works to be constructed and completed in accordance with the contract.

They represent the finished works or structures for which the project was commissioned. The definition of temporary works is in clause 1 (m):

Temporary works means all temporary works of every kind required in or about the construction and completion of the works.

In so far as the conditions of contract are concerned, the 'works' are defined as the combination of the temporary and permanent works (clause 1 (o)).

Design responsibility

The contractor is not generally responsible for the permanent works in terms of their:

- design
- specification

However, there is provision in the conditions of contract to cover permanent works designed by the contractor (clause 7 (6)). This is regarded as contractor-designed construction and, in some circumstances, will form a part of the contract. The contractor is required to exercise all reasonable skill, care and diligence in designing any part of the permanent works. The contractor is also not responsible for sufficiency of the temporary works that have been designed by the engineer. However, the responsibility for the majority of the temporary works will be the contractor's. The provision of temporary works will be greatly influenced by the methods that the contractor chooses to carry out and complete the works.

Contractor responsible for safety of site operations

The contractor must take full responsibility for both the temporary or permanent works during construction. This includes responsibility for all site operations and methods of construction in respect of their:

- adequacy
- stability
- safety

Reference should also be made to the Construction (Design and Management) Regulations 1994 (clause 71). These are considered in Chapter 15.

Contract Agreement (clause 9)

The contractor may be required to complete a Contract Agreement. This will be at the expense of the employer, using the form in the Conditions of Contract.

Performance security (clause 10)

It has become common today for employers to require performance bonds for construction contracts. This is to ensure that the contractor completes the works. The bond provides for a third party, often a bank, to provide the additional funds that may be required to complete a project in the event of the contractor becoming insolvent or for some other reason not completing the works. Where a bond is to be provided, this will be stated in the contract documents, usually in the specification. Under the Conditions of Contract the employer can request such a security for up to 10 per cent of the tender total. This will be required within 28 days of the awarding of a contract to a contractor. The security for the bond must be provided by a body that is approved by the employer. A form of bond is provided in the Appendix to the Conditions of Contract. It is usual for the fees associated with a bond to be paid by the contractor, unless the contract provides otherwise. Since the employer is likely to benefit where a bond is brought into use, the employer is deemed to be a party to the security.

In practice, a contractor may be prohibited from even tendering for a project where it is known either that it will not be possible to arrange a bond, or that it will be prohibitive in terms of its fee. For example, a civil engineering contractor who has traditionally carried out repairs to urban roads and sewers, up to £1m, may be unlikely to obtain a bond (at the appropriate price) for culverting a small river at a cost of £10m. The provision of a bond is not solely restricted to past performance but also the general financial and management capability of a contractor.

Arbitration upon security

The employer is deemed to be a party to the performance bond. Any dispute between the employer and the contractor will be dealt with under clause 66.

Provision and interpretation of information (clause 11)

The employer will be deemed to have made available to the contractor, before the submission of a tender all information on:

• nature of ground, subsoil and hydrological conditions
• pipes and cables in, on or over the ground

This information will have been obtained by the employer through a site investigation prior to the design of the works. It is the contractor's responsibility to correctly interpret this for the purpose of constructing the works. It may also be used where the contractor is responsible for designing any part of the works.

Inspection of site

Before submitting a tender, the contractor will also have inspected and examined the site and its surroundings and used the information described above. The contractor will be deemed to have determined the:

• form and nature of the site, including the ground and subsoil
• extent and nature of the project

- work and materials necessary for constructing and completing the works
- communication and access to the site
- accommodation that may be required
- possible risks and contingencies

Basis and sufficiency of tender

The contractor will therefore formulate the tender on the basis of:

- information supplied to them from the employer
- inspection of the site
- examination of the contract documents
- correctness and sufficiency of the rates and prices in the bills of quantities which will cover all of the contractor's obligations

Adverse physical conditions and artificial obstructions (clause 12)

During the execution of the works the contractor may encounter physical conditions or artificial obstructions. If these could not have been reasonably foreseen by an experienced contractor, a written notice of such should be given to the engineer. These conditions exclude weather conditions and conditions due to bad weather.

The use of this clause on major civil engineering works is common in the support of claims for additional expense. This is due in part because of the nature of civil engineering works and the fact that much of the work is at or below ground level. More sophisticated means of determining the full extent of the ground conditions are constantly being brought in to use. However, it is often not until the contractor has commenced work that the full extent of the site is properly understood. It is the engineer's responsibility to ensure that the information supplied to the contractor during tendering is as correct as possible. The contractor, even an experienced firm, cannot be expected to make allowances or to read between the lines of a site investigation report.

Intention to claim

At the time of writing to the engineer the contractor should state whether a contractual claim is to be made. The claim may be in the form of an additional payment (clause 52) or an extension of time (clause 44). Where it is not possible to provide this information about the adverse physical condition at the time of writing, the engineer should be informed about this possibility as soon as possible.

Measures being taken

The contractor should also provide details to the engineer of any anticipated effects of the adverse physical condition. This will include the measures that the contractor has taken, is taking or will take to deal with the problem. The contractor should also inform the engineer of the:

- estimated costs involved
- extent of anticipated delay
- effects upon the works

Action by engineer

Following the receipt of the notification relating to possible adverse physical conditions, the engineer should consider whether to

- ask the contractor to investigate and report upon possible alternative measures that might be used together with their costs and timing
- give written consent to the measures notified by the contractor
- give the contractor written instructions on how the adverse physical condition is to be dealt with
- suspend the works under clause 40
- issue a variation under clause 51

Conditions reasonably foreseeable

The engineer may decide that the physical conditions or artificial obstructions could have been reasonably foreseen by an experienced contractor. In this case the engineer will inform the contractor as soon as possible in writing of this decision. Where the contractor's notification under clause 12 has resulted in the engineer issuing a variation under clause 52, then the costs of this variation will be added to the contract price.

Delay and extra cost

Where the engineer considers that such conditions or obstructions could not have been foreseen by an experienced contractor then an additional payment, including a reasonable amount for profit and/or an extension of time should be granted to the contractor. The engineer will notify the contractor in writing with a copy given to the employer.

Works to be to satisfaction of the engineer (clause 13)

The contractor is to construct and complete the works in strict accordance with the contract and to the satisfaction of the engineer. Only if the work is illegal or physically impossible to carry out, will the contractor have any redress for non-compliance. The contractor must comply with all the engineer's instructions connected with the project. The contractor must take instructions only from the engineer or, subject to certain limitations (clause 2), the engineer's representative. The engineer has wide powers, but if the contractor feels that these are being exceeded, the matter should be brought to the attention of the engineer (clause 2 (7)).

Mode and manner of construction

The materials, equipment and labour provided by the contractor (clause 8) and the mode, manner and speed of construction of the works are of a kind and conducted in a manner acceptable to the engineer.

Delay and extra cost

The engineer's instructions can sometimes cause delay or disruption to the contractor's arrangements or method to be used for construction purposes. Where the contractor incurs costs beyond those reasonably foreseen by an experienced contractor at the time of tender, the engineer will take these into account when determining an extension of time under clause 44. The contractor will be paid for such delays or disruptions in accordance with clauses 52 (Variations) and 60 (Certificates and payments). Profit will be added to the costs of additional temporary or permanent work. If the delays or disruptions are partially due to the contractor, the amount of delay and costs involved will be adjusted accordingly. Where such instructions require a variation, the same shall be assumed to have been given pursuant to clause 51.

Removal of contractor's employees (clause 16)

The contractor must employ only those who are careful, skilled and experienced in their several trades and callings.

The engineer can object to any person employed on the works and require the contractor to remove them from site. This must not be done unreasonably or to the specific annoyance of the contractor. The contractor should therefore have been previously warned of this impending situation, and given an opportunity to take appropriate corrective action. It is, thankfully, an uncommon occurrence. However, it may arise in circumstances where persistently bad workmanship arises or where a member of the contractor's staff has consistently failed to carry out the engineer's instructions as requested.

If, in the opinion of the engineer, a person is guilty of:

- misconduct
- incompetence
- negligence
- failing in the performance of duties
- failing to conform with particular provisions regarding safety
- persisting in conduct which is prejudicial to safety or health

then such a person must not be re-employed upon the works without the permission of the engineer. For example, where an employee is dismissed by the contractor for one of the above reasons, that same person would not be able to work again on that project, in the services of a subcontractor.

Setting out (clause 17)

The engineer is responsible for providing the contractor with all the information necessary for the contractor to set out the works. This information should comprise properly dimensioned drawings showing line, level and location. Delays in the presentation of the information from the engineer could result in an extension of time.

The setting out of the works is the responsibility of the contractor. This includes the provision of all necessary instruments, appliances (e.g. profile boards) and labour. The setting out involves the correctness of the:

• position
• levels
• dimensions
• alignment

If errors in setting out are subsequently discovered they must usually be rectified at the contractor's expense to the satisfaction of the engineer. Where an error occurs due to incorrect information being supplied by the engineer or the engineer's representative, the cost of rectifying this is borne by the employer. One possible problem which could arise from inaccurate setting out of the works would be trespass on the land of an adjoining owner. With the consent of the employer, the engineer may instruct that errors will not be corrected and an appropriate adjustment for such errors will be made from the contract sum.

The engineer may frequently check the contractor's setting out, or at least the key positions. However, errors in the setting out made by the contractor are not relieved by this action. It is also the contractor's responsibility to protect and preserve all bench marks, sight rails, pegs and other things used in the setting out of the works.

Boreholes and exploratory excavation (clause 18)

The engineer can at any time during the progress of the works instruct the contractor to make boreholes or carry out exploratory excavations. Where a provisional or prime cost sum (see clauses 58–59) has been included in the bill of quantities, the costs of such work will be offset against these items. Where the bill of quantities does not include such items, the instruction shall be deemed to be a variation under clause 51.

Safety and security (clause 19)

The contractor should, throughout the progress of the works, have full regard for the safety of all people who are entitled to be upon the site. The contractor should keep the site, which is under the control of the contractor, and works which are not completed or occupied by the employer in an orderly state appropriate to the avoidance of danger. The contractor should also provide and maintain:

- lights
- guards
- fencing
- warning signs
- watching

These are to be available when and where necessary or required by the engineer or engineer's representative or competent statutory authority, such as the Health and Safety Executive. They are to be provided for the protection of the works and for the safety and convenience of the public and others.

Employer's responsibilities

The employer may employ other contractors or use a form of direct labour for certain aspects of the work (clause 31). In these cases the contractor has little jurisdiction over such firms. It is the employer's responsibility in these circumstances to ensure that the same regard for safety and the avoidance of danger is adopted by such firms and workmen.

Care of the works (clause 20)

The contractor is to take full responsibility for the care of the works and the materials, plant and equipment that are to be incorporated in the works from the contract commencement date until the date of the issue of the certificate of substantial completion for the whole of the works. At this point the responsibility for the care of the works passes to the employer. If the engineer issues a certificate of substantial completion for any section of the permanent works, the contractor will cease to be responsible for the care of that section of the works. The contractor is responsible for the care of any outstanding work during the defects correction period or until such work is completed.

Excepted risks

There are a number of excepted risks for which the contractor is not liable for loss and damage. These include:

- use or occupation by the employer and those who are directly responsible to the employer
- any fault, defect, error or omission in the design of the works, other than in contractor-designed construction
- riot, war, invasion, act of foreign enemies or hostilities
- civil war, rebellion, revolution, insurrection or military or usurped power
- ionising radiations or contamination by radioactivity from any nuclear fuel or from nuclear waste from the combustion of nuclear fuel, radioactive toxic explosive or other hazardous properties of any nuclear assembly or nuclear component
- pressure waves caused by aircraft or other aerial devices travelling at sonic or supersonic speeds

Rectification of loss or damage

The contractor is responsible for making good any loss or damage, other than excepted risks described above, to the works or a section of the works and to materials, plant or equipment to be incorporated in the works. The contractor is also responsible for making good any loss or damage that might occur during the compliance with the obligations under clauses 49 and 50 (outstanding work and defects).

The contractor is not liable for any loss or damage arising as a result of one of the excepted risks, listed above. The engineer may require the contractor to rectify such loss or damage, but this will be at the expense of the employer.

Where loss or damage is a combination of both excepted risks and the contractor's own negligence, then the costs of rectification will be apportioned between the contractor and the employer.

Insurance of the works (clause 21)

The contractor must insure the works in the joint names of the contractor and employer. This will include the materials, plant and equipment that are to be incorporated within the works. The extent of the insurance cover is to be the full replacement costs, plus an additional 10 per cent to allow for incidental costs. These incidental costs might include demolition and removal of debris, professional fees and extra costs incidental to the insurance work. The principle of the insurance clause is that the employer should be left in the same financial position as would have been the case prior to an insurance claim arising.

Extent of cover

The insurance required for the contract must cover any loss or damage other than that arising from the excepted risks that have been defined in clause

20 (2). The period of cover runs from the works commencement date until the issue of the certificate of substantial completion.

The insurance must also extend to cover any loss or damage arising during the defects correction period from a cause occurring prior to the issue of the certificate of substantial completion. The insurance must also cover the work carried out by the contractor in completing any outstanding work and defects under clauses 49 and 50.

Any amounts that are not insured or cannot be recovered from the insurers, e.g. excess amounts, are to be borne by either the contractor or the employer in accordance with their respective responsibilities under clause 20.

Damages to persons and property (clause 22)

The contractor must, unless the contract provides otherwise, indemnify the employer in respect of death or injury to any person and loss or damage to property, other than the works, resulting from executing the works and remedying any subsequent defects.

Exceptions

There are exceptions within the contract to this indemnity by the contractor. These may result from the unavoidable construction of works in accordance with the contract. These are as follows and are borne by the employer:

- damage to crops on the site
- the use or occupation of the land for construction purposes which might unavoidably result in interference with a right of way or other easement
- the right of the employer to construct the works or any part, on, over, under, in or through any land
- damage which is unavoidable as a result of constructing the works in accordance with the contract
- death or injury to persons or damage to property committed by the employer, an agent, servant or contractor employed directly by the employer

Indemnity by employer

The employer must indemnify the contractor against all possible claims that might arise out of the above events.

Shared responsibility

Death, injury, loss or damage may be due to an act of negligence on the part of the employer or persons for whom the employer is separately and directly responsible. Under these circumstances the liability of the contractor will be reduced in proportion to such negligence.

The contractor, or those for whom the contractor is directly responsible, e.g. subcontractors, may have contributed towards death, injury, loss or damage under one of the exceptions listed above. The employer's liability to indemnify the contractor is similarly reduced.

Third party insurance (clause 23)

The contractor will insure in the joint names of the contractor and employer. This will not necessarily limit either the contractor's or the employer's liability. The third party insurance excludes the employees of the contractor, subcontractors and the works. All of these are insured for separately.

Cross liability clause

The insurance policy will include a cross liability clause. The insurance can apply to the contractor or employer separately.

Amount of insurance

The amount of the insurance must be for at least the amount stated in the Appendix to the form of tender (see appendix part 2).

Accident or injury to workpeople (clause 24)

Accidents and injuries may occur on site to the contractor's or a subcontractor's employee. The employer is not liable for any damages or compensation that may arise because of such events. The contractor must therefore indemnify the employer against possible damages or compensation and their respective claims, costs, charges or expenses. Only where such accidents or injuries result directly from the employer or an agent or servant will the employer be liable.

Evidence and terms of insurance (clause 25)

The contractor must, prior to the start of the project, i.e. works commencement date, provide evidence to the employer that the required insurances have been brought into effect. Policies may be required for inspection by the employer. The terms of the insurances will be subject to reasonable approval by the employer. The contractor must, on request, provide receipts for the payment of current insurance policies.

Excesses

Any excesses on the policies of insurance must be stated on the Appendix to the form of tender (see appendix part 2).

Remedy on contractor's failure to insure

Where the contractor defaults in the payment of insurances or cannot provide the necessary evidence that such policies have been paid, the employer can take out insurance on behalf of the contractor. Any premiums that are required will then be deducted from amounts due to the contractor.

Compliance with policy conditions

Both the employer and the contractor must comply with all the conditions that are laid down in the insurance policies. Where either party fails to comply, the other party is indemnified against all losses or failures that might arise from such failures.

Giving notices and payment of fees (clause 26)

The contractor is responsible for paying all notices, fees and charges. These may be required by an Act of Parliament, a regulation or a bye-law. They may relate to the construction or completion of the works. They may affect public bodies or companies whose property or rights are affected by the works.

Repayment by employer

The employer will repay all of the required sums that have been properly certified by the engineer in respect of such fees, rates and taxes. These may apply to temporary structures as well as the permanent works. They include anything that is exclusively for the purpose of the works.

Contractor to conform with statutes etc.

The contractor must comply with any general or local Act of Parliament, regulations or bye-law. These may have been approved by a local or statutory authority. The contractor must keep the employer indemnified against all penalties and liabilities of any kind. The exceptions to this include:

- breaches which are unavoidable as a result of complying with the contract
- instructions given by the engineer which do not conform with such an Act, regulations or bye-laws. (The engineer must issue a variation under clause 51 in order to conform to such an Act, regulation or bye-law.)

- the contractor is not responsible for obtaining planning permission in respect of any permanent works or temporary works that have been designed by the engineer. The employer warrants that such permissions have or will be obtained.

New Roads and Street Works Act 1991 – Definitions (clause 27)

In this clause the Act means the New Roads and Street Works Act 1991 and any statutory modification or re-enactment.

The employer is the licensee for the purpose of obtaining any licence under the Act for the permanent works. For all other purposes the undertaker under the licence is the contractor. The interpretation of this clause is the same as that intended in the Act.

Licences

It is the employer's responsibility to obtain any street works licence that may be required for carrying out the permanent works. The contractor is to be supplied with a copy of the licence, together with details of any conditions or limitations that might be imposed. Any condition or limitation that is imposed after the award of the contract will be deemed to be an instruction under clause 13.

Notices

It is the contractor's responsibility to give the relevant authority notice under this clause. A copy is to be given to the employer.

Patent rights (clause 28)

The contractor must indemnify the employer against any claims or proceedings that may arise due to an infringement of patent rights. These include any design trademark or other protected right in respect of:

- contractor's equipment used on the works
- materials, plant or equipment to be incorporated in the works

The employer must indemnify the contractor against any claims that might arise resulting from the design of specification in this respect.

Royalties

The contractor is responsible for the payment of all tonnage and other royalties, rent, etc. for getting stone, sand, clay or other materials required for the works.

Interference with traffic and adjoining properties (clause 29)

The contractor should carry out the works so as not to interfere unnecessarily or improperly with the public. This includes access to public and private roads, footpaths or property. The contractor must indemnify the employer in respect of all possible claims arising from such matters.

Noise disturbance and pollution

All of the works are to be carried out without unreasonable noise, disturbance or other pollution.

Indemnity by contractor

Noise, disturbance or other pollution is often unavoidable in construction works. The contractor must indemnify the employer against any liability for damage and against any claims that might arise in this context.

Indemnity by employer

Similarly, the employer is to indemnify the contractor in respect of damage and claims that might occur.

Avoidance of damage to highways (clause 30)

The contractor is to use every reasonable means to protect highways and bridges on route to the site. The contractor must not subject these to extraordinary traffic within the meaning of the Highways Act 1980. The contractor must ensure that subcontractors also conform to this requirement. The contractor should be careful in selecting routes and in using appropriate vehicles to move the contractor's plant and equipment, materials and prefabricated components. Care should be exercised at all times and no unnecessary damage or injury should be made to highways and bridges.

Transport of contractor's equipment

The contractor is responsible for the strengthening of bridges that may be required during the transportation of contractor's equipment and temporary works. In some cases it may be necessary to alter or improve a highway for these purposes. The cost of this is to be borne by the contractor, who must also pay any other charges that may be involved. The contractor must also indemnify the employer in this respect and negotiate and pay all claims should these arise.

Transport of materials

The contractor should notify the engineer as soon as any damage occurs to bridges or highways due to the transportation of materials or manufactured articles. Normally the hauliers of such items are required under an Act of Parliament or statutory instrument to have indemnified the highway authority against the possibility of such damage occurring.

In other cases the employer will negotiate and settle any sums that may be due. Where, for example, the work results from a variation, the contractor should be repaid the costs incurred under this clause.

Facilities for other contractors (clause 31)

The contractor will appreciate that he does not have exclusive possession of the site. The contractor must therefore allow all reasonable facilities for other contractors who may be employed independently of the contract by the employer. Such access to the site will also apply to other properly authorised organisations or statutory bodies who may be employed on or near the site but which are not a part of the contract either.

Delays and extra cost

The compliance with this clause may involve the contractor in a delay or additional cost. Where this could not have been foreseen by an experienced contractor at the time of tender, the engineer must take this into account when determining any extension of time. The extension of time will be dealt with under clause 44 and any cost under clause 52 (4). Any agreed reasonable costs involved will be paid under clause 60, together with profit in respect of any additional permanent or temporary work.

Fossils, etc. (clause 32)

All fossils, coins, articles of value or antiquity and structures or other remains of geological or archaeological interest that are discovered on the site are the property of the employer. The contractor must take reasonable precautions against their possible damage or theft. Upon discovering such items, the contractor must immediately inform the engineer. The engineer must then give instructions, on behalf of the employer, to the contractor. The additional work involved will be at the expense of the employer.

Clearance of site on completion (clause 33)

Upon completion of the works, the contractor must clear away and remove from the site:

- contractor's equipment
- surplus material
- rubbish
- temporary works

The whole of the site and permanent works must be left in a clean and workmanlike condition to the satisfaction of the engineer.

Returns of labour and contractor's equipment (clause 35)

The contractor shall, if required by the engineer, provide details and totals of the different kinds of labour being used on the site at different times. Similarly, the engineer may request details of the contractor's plant and equipment. The clause also applies to the contractor's subcontractors employed on the works.

CHAPTER 15

WORKMANSHIP AND MATERIALS

Introduction

The clauses within the ICE Conditions of Contract that refer to the quality of materials and standards of workmanship to be expected during construction are:

- workmanship and materials (clauses 36–40)
- the Construction (Design and Management) Regulations 1994 (clause 71)

The specification provides most of the information relevant to the project. This document will seek to identify the actual quality of materials and standards of workmanship expected on a particular project.

The combination of ensuring that the contractor complies with the statutory regulations, and the specification of the works described, should result in a civil engineering project that achieves the desired quality and standards. There is, in addition, provision for the engineer to inspect the works before, during and after construction. The Conditions also allow for the precautionary inspection of goods and materials in the workshops off site, should this be so desired. The engineer is to be given the opportunity of inspecting manufacturing process of the different materials and equipment.

The engineer is able to assess the standards of the materials and workmanship of the various components against those laid down in British Standards or Codes of Practice. On some occasions the standards indicated will be higher to meet specification requirements. In some cases they may be lower or relaxed, where an aspect of the work being carried out is not of critical importance.

The contractor should seek to ensure that competent tradesmen are employed to do the work and that they are properly supervised. A contractor's representative or agent must be available on the site to receive oral and written instructions from the engineer. The agent will also be responsible for the daily site management of the project.

There is provision for uncovering work for inspection where the engineer feels that this is necessary. Whilst the work in progress can be inspected at any time, the engineer has the power to open completed work for inspection and testing to ensure that it complies with the specification. Under these circumstances, the principle is that where the work proves to be satisfactory the employer must bear the costs involved. Where the work fails to comply with the specification, the costs of opening up, correcting and reinstating must be borne by the contractor.

In carrying out the works the contractor must comply with all statutory obligations, notices, fees and charges (clause 26). The contractor must comply with and give all notices required by any Act of Parliament, statutory instrument, regulation or bye-law of any local authority or statutory undertaker which has any jurisdiction with regard to the works. The contractor, however, will not be liable for compliance with this requirement where the works themselves do not comply with these requirements. This may be because a particular regulation or bye-law has been relaxed for the duration of the project.

The contractor should not deliberately and knowingly execute work that will contravene these regulations, and will therefore require future modification. On the other hand, the contractor is not a watchdog to ensure that engineers carry out their duties properly and efficiently. If the contractor finds any divergence between the statutory requirements and any of the contract documents or engineer's instructions, then the engineer should immediately be informed in writing. The engineer should be made aware of the possible discrepancy and the contractor should be given instructions on which document to follow. There is no hierarchy of documentation under the ICE Conditions of Contract. Where the instruction requires the work to be varied, this should be treated in the same way as other variations, in accordance with clause 51.

In some circumstances it may be necessary for the contractor to comply urgently with a statutory regulation – for example, where an existing structure on the site is in danger of collapsing or where health is in danger. This may require the contractor to supply materials or execute work prior to receiving instructions from the engineer. The contractor should do that which is reasonably necessary to secure immediate compliance with the statutory requirement. The contractor must also inform the engineer of this action. The materials and work executed in these circumstances will then be treated as a variation under clause 51. The engineer has power to deal with urgent works of this type under clause 62.

The contractor must pay and indemnify the employer against liability in respect of any fees, charges, rates, taxes, etc. These may be required under an Act of Parliament, regulation or bye-law.

Clause 71 covers the provisions of the Construction (Design and Management) Regulations 1994. The employer must ensure that the planning supervisor, defined as a chartered engineer, carries out the relevant duties under the CDM Regulations.

The engineer is responsible for providing the contractor with all the information necessary for the contractor to set out the works at ground level.

This should comprise properly dimensioned drawings and levels and the information identified in clause 17. The actual setting out of the works is entirely the responsibility of the contractor. If mistakes are made during this process, the contractor must correct them without cost or charge to the employer. Delays in the receipt of the information from the engineer could result in an extension of time. One possible problem arising from the inaccurate setting out of the works could be trespass on the land of an adjoining owner. In this situation the contractor would be liable to indemnify the employer under clause 22. With the consent of the employer, the engineer may instruct that such errors will not be amended and an appropriate deduction for such errors will be made from the contract sum.

WORKMANSHIP AND MATERIALS (CLAUSES 36–40)

The quality of workmanship and materials is tested as detailed, with remedies and solutions:

Quality of materials and workmanship and tests (clause 36)

The quality of materials and standards of workmanship on a project are:

- of the respective kinds described in the contract
- in accordance with the engineer's instructions
- subject to tests as the engineer may direct, whether in the place of manufacture or fabrication, on site, or elsewhere, as specified in the contract

The contractor is required to provide:

- assistance
- instruments
- machines
- labour
- materials

normally required for examining, measuring, and testing any of the work or materials. The contractor will be requested to supply samples of materials before their incorporation within the works. These samples will be selected by the engineer as required.

Cost of samples

The samples of materials for testing purposes will have been described in the contract (specification) and will therefore be deemed to have been included in the contractor's tender. Where these are not described or where samples of other materials are required, the costs of these will be borne by the employer, i.e. the contractor will be paid an additional sum.

Cost of tests

The same principle will normally be applied to the costs of making any test. The contract documents, normally the specification, need to identify clearly the tests that will be required. They must be described in sufficient detail to allow the contractor to be able to price them at the tender stage. The engineer may also wish to apply other tests on the materials and workmanship in order to satisfy the requirements of the contract. The costs of these tests and tests that were unclear at the time of tender will normally be borne by the employer. However, if such tests indicate that the materials are not in accordance with the provisions of the contract, the costs involved will be borne by the contractor.

Access to site (clause 37)

The engineer, or a representative, is allowed reasonable access to the site at any time. The engineer is also to be allowed access to:

- workshops
- any places where work for the contract is being prepared
- materials (and component) manufacturers
- machinery manufacturers

The contractor is to offer every facility and assistance in obtaining such access or the right to access.

Work covered up (clause 38)

Permanent work should not normally be covered up until it has been inspected and approved by the engineer. The contractor must give the engineer reasonable notice to allow any examination and measurements to be taken – for example, on foundations before the superstructure is constructed on them. The engineer must inspect the work without unreasonable delay or advise the contractor that this will not be required.

Uncovering and making openings

The contractor is required to uncover any part of the works at any time and to make good to the satisfaction of the engineer. Where the work executed is in accordance with the contract, the cost of uncovering and making good or reinstating will be at the employer's expense. This is providing that the contractor informed the engineer that the work was ready for covering up.

Removal of unsatisfactory work and materials (clause 39)

During the progress of the works, the engineer has the authority at any time to instruct the contractor, in writing, to:

* remove from site materials, that in the opinion of the engineer, are not in accordance with the contract
* substitute materials
* remove and re-execute work, that is not in accordance with the contract, in respect of materials and workmanship or contractor's designed work

Default of contractor in compliance

Where the contractor fails to carry out such instructions within a reasonable period of time, other firms can be employed for this purpose. The costs involved will then be deducted from amounts due to the contractor.

Failure to disapprove

The engineer is able to take action under this clause at any time. A failure to disapprove any work or materials will not prejudice the power of the engineer to take action subsequently.

Suspension of works (clause 40)

The engineer has the power to suspend the progress of the construction works in whole or in part. During such a suspension the contractor must ensure that the works are properly protected and secure. The contractor will be paid for any extra work involved under clause 52 (4) and in accordance with clause 60. Suspension of the works may be occur because of:

* some peculiar aspect of the project
* weather conditions
* a default on the part of the contractor
* considerations of safety
* excepted risks under clause 20 (2)

When dealing with matters of an extension of time under clause 44, the aspect of suspension will also be taken into account. If the suspension of the works was an integral part of the contract or is due to a default on the part of the contractor, then an extension of time will not be considered.

Suspension lasting more than three months

Unless the suspension of the works is part of the contract, the contractor, within a period of 3 months from the date of suspension, may request to proceed with the works. This request must be in writing to the engineer, with a reply expected within a further 28 days. Where this permission is not granted, the contractor must again write to the engineer for clarification. If the suspension relates only to a part of the works, then this will then be assumed to be an omission of this part from the contract under clause 51. Where it affects the whole of the works, the contractor can assume that the works have been abandoned by the employer.

Construction (Design and Management) Regulations 1994 (CDM) (clause 71)

The construction industry is a dangerous environment. It has a poor health and safety record. Serious injury and death happen far too frequently as a result of construction activities, especially on site and affect not only construction workers but also members of the general public. Improving the management process is essential in helping to prevent accidents and ill health in the industry.

Different governments have initiated measures aimed at reducing accidents on construction sites, most notably the Health and Safety at Work Act, 1974, which applies to all industries. However, it is recognised that some industries are more dangerous than others. The data shown in Tables 15.1 and 15.2 help to illustrate this point.

Falling from a height was the most common cause of fatality in the industry. Being struck by a moving vehicle was the next most common. Lifting and carrying accounted for nearly a quarter of all reported injuries in 1992/93.

The CDM Regulations, which became operative in 1995, apply to all construction projects and everyone associated with their design and construction. The Regulations are about the management of health and safety and place new duties on employers, planning supervisors, designers and contractors to

Table 15.1 Injuries to employees: construction and all industries

Type of accident	Rates per 100,000 employees			
	1986/87	1988/89	1990/91	1993/94
Fatal and major				
Construction	293	296	291	233
All industries	101	94	92	79
All reported injuries				
Construction	1995	1928	1907	1416
All industries	862	842	818	710

Source: Department of the Environment

Table 15.2 Injuries to employees in the construction industry 1992/93

Cause of injury	Fatal	Major	Total
Contact with moving machinery	2	71	73
Struck by flying object	7	310	317
Struck by moving vehicle	14	90	104
Slip, trip or fall	1	366	367
Fall from height	27	808	835
Trapped by collapse	6	76	82
Exposure to harmful substance	4	41	45
Contact with electricity	8	63	71
Other	0	231	231
Totals	69	2056	2125

Source: Department of the Environment

plan, co-ordinate and manage health and safety throughout all stages of a construction project. Anyone who appoints a designer or contractor has to ensure that they are competent for the work and will allocate adequate resources for health and safety. There are five key parties, firms or individuals, who are involved and each has specific duties to perform.

- Employers: These should be satisfied that only competent people are appointed as planning supervisor and principal contractor. This also applies when making arrangements for the appointment of designers and contractors. Employers should ensure that sufficient resources have been allocated to ensure that the project can be carried out safely.
- Designers: These should ensure that structures are designed to minimise risks to health and safety while they are being constructed and maintained. Where risks are unavoidable, adequate information must be provided. The design includes both drawn and written information, e.g. specifications.
- Planning supervisors: These have the overall responsibility for co-ordinating the health and safety aspects of the design and planning phase and for the early stages of health and safety plan and the health and safety file.
- Principal contractors: These should take account of health and safety issues when preparing and presenting tenders or similar documents. The principal contractors have to develop health and safety plans and co-ordinate the activities of all contractors to ensure that each complies with the appropriate health and safety legislation and regulations. The principal contractors also have a duty to check on the provision of information and training for employees and to consult with employees and the self-employed on health and safety matters.
- Contractors and the self-employed: These should co-operate with the principal contractor and provide relevant information on the health and safety

risks created by their work and how they will be controlled. Contractors also have a duty to provide other information to the principal contractor and to employees. The self-employed have similar duties to contractors.

Employees on construction sites should be better informed and have the opportunity to become more involved in health and safety matters.

The new procedures will cost the industry and hence its employers more money and time to implement and monitor, at least initially. However, these procedures may in the long term help to reduce costs through developing better construction practices. They should help to save lives and reduce accidents and the disruption accidents sometimes cause to work on site. They will reinforce the need for co-ordinating and managing health and safety from inception to completion and during the use of the completed project.

CHAPTER 16

TIME

Introduction

Several clauses in the ICE Conditions of Contract are of relevance and importance in respect of the time factor in the construction of civil engineering projects. These include:

- commencement time and delays (clauses 41–46)
- liquidated damages for delay (clause 47)
- certificate of substantial completion (clause 48)
- outstanding work and defects (clause 49–50)
- frustration (clause 64)

This group of clauses describes the start and finish of the project and the implications associated with these dates. The difference between these is, of course, the contract period. The time for completion is stated in the Appendix to the form of tender and is the date when the certificate of substantial completion is due to be issued. Should the contractor fail to complete the works by this date, the engineer must consider:

- awarding the contractor an extension of time, or
- levying liquidated damages for non-completion.

The time for completion cannot occur prior to the date originally anticipated in the contract and stated in the Appendix, unless the contractor agrees to such a revision. The same principle applies where the contract is to be executed in sections rather than as a whole. As far as time is concerned, each section of the works, where appropriate, is treated almost as a separate project. Upon the issue of the certificate of substantial completion the contractor is relieved of most of the contractual obligations. However, the contractor is still responsible for defects that may arise from substandard materials or bad workmanship. This responsibility continues until the end of the defects correction period. The ICE Conditions do not recommend a period of time as is common with some other forms of contract. Six months is a typical period, but the complexity of civil engineering works frequently means that this period is extended to at least 12 months. Currently there is some debate on whether this responsibility should extend further – for some aspects, such as structural work, up to 10 years.

In some extreme circumstances the works may be frustrated, perhaps indefinitely, and clause 64 allows for this eventuality.

COMMENCEMENT TIME AND DELAYS (CLAUSES 41–46)

The Scheduling of the project and circumstances affecting the Schedule are described:

Works commencement date (clause 41)

Works commencement date

This will usually be stated in the Appendix (Part 1) to the form of tender alongside the time for completion (clause 43). Where a date is not included, then the commencement date will be within 28 days of the award of the contract or any other dates agreed between the parties, i.e. the employer and the contractor.

Start of works

The contractor must commence work on the project as soon as possible after this date, although no specific period is indicated in the Conditions. The contractor must with due expedition and without delay carry out the works in accordance with the terms of the contract.

Possession of site and access (clause 42)

The contract may also prescribe matters relating to the possession and access to the site. For example, the employer may require the project to be carried out in a particular order or sequencing, or may restrict the contractor's access to the site to a specified number of entry and exit points. In the case of motorway projects, the contractor is unlikely to be allowed to gain access to the site from every existing road that crosses the new motorway. The contract may therefore prescribe:

- the extent and portions of the site of which the contractor is to be given possession from time to time (full or partial possession of the site)
- the order in which such portions of the site shall be made available to the contractor (specified order in which the different parts of the site are to be made available to the contractor)
- the availability and the nature of access which is to be provided by the employer (points of access and egress to the site)

- the order in which the works are to be constructed (specified order for completing the works)

The employer must release a sufficient portion of the site to enable the contractor to commence and proceed with the works. Where the entire site has not been released at the commencement date, further parts of the site will need to be released to enable the contractor to complete the works. This will be in accordance with the agreed programme that has been accepted by the engineer under clause 14.

Failure to give possession

If the employer does not give possession of the site to the contractor at the agreed times, then the contractor may be entitled to:

- an extension of the contract period (extension of time) under clause 44
- additional costs, which can also include profit

Access and facilities provided by the contractor

The contractor is responsible for all of the costs associated with access to the site. These costs may need to include the costs of any additional facilities outside the perimeter of the site which may be necessary for the contractor to complete the works in a satisfactory manner. For example, they may include temporary works, such as traffic signals, barriers, etc. to protect third parties.

Time for completion (clause 43)

The time specified for completion of the works (or a section of the works) is stated in clause 48 and in the Appendix to the form of tender. Completion is usually interpreted as substantially complete, which implies fit for the use intended by the employer. In practice, the certificate of substantial completion (clause 48) will be issued when the engineer feels that the works have reached this stage.

Extension of time for completion (clause 44)

Where the works are not completed by the date stated in the Appendix to the form of tender, the engineer will need to assess the reasons for the delay in order to determine who is responsible. If the delay is the fault of the contractor, it is usual to apply some form of liquidated damages (clause 47). Where the delay has been caused by the employer (or the engineer acting on behalf of the employer), the contractor may be entitled to an extension of time, i.e. the time for completion is lengthened to take into account the delay. The granting of an extension of time to the contractor may be a mechanism to reduce the liability for paying liquidated damages. An extension of time may or may not result in additional payments to the contractor.

Where a delay to the contract programme arises due to one of the following reasons, then it is the contractor's responsibility, within 28 days, to provide the engineer with the full particulars and a justification of the period of extension to be claimed. This is to enable the engineer to investigate properly the reasons for the delay. A contractor may be entitled to an extension of time for

- variations ordered under clause 51 (1)
- increased quantities referred to in clause 51 (4)
- any cause of delay referred to in the contract conditions
- exceptional adverse weather conditions
- any other special circumstances

Assessment of delay

The contractor must be notified of the engineer's decision in writing. The engineer also has the option of awarding the contractor an extension of time, where the engineer believes that one of the above events has delayed the works.

Interim grant of extension of time

In the first instance it is usual for the engineer to award an interim extension of time that will be either confirmed or refuted at a later date.

Assessment at due date for completion

Any extension of time must be determined within 14 days of the time for completion or an extended date where this has already been agreed. Where the engineer considers that the contractor is not entitled to an extension of time, both the contractor and employer should be informed of this decision.

Final determination of extension

Once an extension of time has been agreed by the engineer it cannot be decreased unless the circumstances change. Normally the time for completion cannot be brought forward without the agreement of the contractor, even where variations may have reduced the work content of the project.

In practice, delays to the contract programme are frequently due to a variety of reasons, that are often a combination of factors on the part of both the employer and the contractor. It is the engineer's responsibility to identify the reasons why a delay has occurred and to apportion the ensuing financial claims that may be made by either party.

Night and Sunday work (clause 45)

The project is assumed to be capable of completion during the daylight hours between Monday and Saturday. In some cases it may be customary to carry out work outside these normal working hours or this may also be specified within the terms of the contract. Work during the night or on Sundays requires the written permission of the engineer. This permission should not be unreasonably withheld (clause 46 (2)). In other cases it may unavoidable or absolutely necessary for the saving of life or property or for the safety of the works. In these circumstances it is also necessary to obtain permission in writing from the engineer. The contractor's tender is deemed to have been calculated on these assumptions. Whilst some civil engineering works are carried out at unsocial times, contractors must not assume in their tenders that this will be approved unless the contract documentation assumes the contrary.

Rate of progress (clause 46)

The contractor must make regular progress with the works in line with the agreed contract programme. Where such progress is not being maintained the engineer must notify the contractor in writing. The contractor must then respond to the engineer with proposals to rectify the lack of progress.

Provision exists within the conditions of contract for accelerated completion. The contractor can be requested by the employer or engineer to complete the works within a shorter period of time than that which was originally envisaged. The conditions allow for special terms and conditions of payment to be agreed under these circumstances.

Liquidated damages for delay (clause 47)

Liquidated damages for delay in substantial completion of the whole of the works

Liquidated damages are defined in Chapter 1. They should represent a genuine pre-estimate of loss where the works remain uncompleted within the time period specified by clause 43. Where an extension of time has been granted under clause 44, the damages due are calculated from this date. They are usually expressed in an amount per week or per day as the case may be.

Liquidated damages for delay in substantial completion where the whole of the works is divided into sections

Where the project has been divided into sections, the principles described above apply to each section of the works. The Appendix to the form of tender must include an amount in respect of each section of the works. Where appropriate, the same principles regarding an extension of time, will be applied to each individual section. Where circumstances dictate, liquidated damages in respect of two or more sections of the works run concurrently.

Damages not a penalty

If damages can be shown to represent an excessive amount they will be deemed to be a penalty and be unenforceable. This does not mean that the employer will not receive any damages for delays in completion of the project, but that the appropriate amounts will be fixed by the courts.

Limitation of liquidated damages

Where no amounts are included in the Appendix to the form of tender or an amount has been omitted, damages will not be payable by the contractor.

Recovery and reimbursement of liquidated damages

Liquidated damages are normally deducted from amounts due under interim payments to the contractor. Alternatively, the employer can send the contractor a separate invoice for the amount due.

The contractor may be subsequently granted an extension of time for the delay. In these circumstances, any liquidated damages that have been paid will be refunded to the contractor. Under these circumstances, the contractor is also entitled to receive interest at compound rates on the same basis as that described in clause 60.

Intervention of variations etc.

If, after liquidated damages have become payable in respect of any part of the works, the engineer issues a variation under clause 51, or adverse physical conditions or artificial obstructions within the meaning of clause 12 occur, then the employer's entitlement to liquidated damages becomes suspended. Such suspension does not invalidate the employer's entitlement to liquidated damages.

Certificate of substantial completion (clause 48)

Notification of substantial completion

This notice is issued to the contractor and the employer when the works are completed, but prior to the contract maintenance period or at the commencement of the defects correction period. It is generally issued for the whole works, though it can be issued in respect of a part of the works where this has been identified in the Appendix to the form of tender (Appendix part 2).

The works must be substantially complete and have satisfactorily passed any final test that may have been prescribed in the contract. Usually the contractor will notify the engineer in writing that the project has reached completion. The terminology 'substantial completion' is, of course, open to some interpretation. In practice, the certificate is usually issued even though minor items of work may still need to be carried out within the terms of the contract. This does not imply that such items of work are unimportant or irrelevant. The notice should be accompanied by an undertaking to finish the remaining work in accordance with clause 49 (1). The date of completion often coincides with the handover date to the employer and this consideration may prompt the issue of the notice even though the works are only 'almost' complete.

Certification of substantial completion

The engineer, must within 21 days of the date of delivery of such a notice, either:

- issue to the contractor and employer a certificate of substantial completion stating the date when the works are substantially complete
- give instructions to the contractor specifying the work which needs to be completed prior to the issue of such a certificate

As soon as the contractor complies with these instructions, the engineer must issue the certificate within 21 days.

Premature use by employer

Where an employer wishes to occupy or use the project, even though it remains unfinished, then the contractor can request the issue of the certificate of substantial completion. Where the engineer considers that any part of the works is substantially complete and has passed a final test prescribed by the contract, a certificate of substantial completion may also be issued.

Substantial completion of other parts of the works

Where the engineer considers that any part of the works has been substantially completed and passed a final test, a certificate of substantial completion in respect of that part of the works can be issued. This will precede the certificate for the whole of the works.

Reinstatement of ground

A certificate of substantial completion for a section of the works, issued before completion of the whole, is not deemed to certify any ground or surfaces still requiring reinstatement.

OUTSTANDING WORK AND DEFECTS (CLAUSES 49–50)

These are covered by the following clauses as detailed:

Work outstanding (clause 49)

The work that remains outstanding at the issue of the certificate of substantial completion must be completed within the time specified by the engineer or agreed between the employer and contractor. Where this is not specified, the outstanding work should be completed as soon as possible and during the defects correction period.

Execution of work of repair, etc.

Notification of the need to make good any defects, of whatever nature, is given by the engineer to the contractor in writing. This should be completed during the defects correction period or within 14 days of its expiry as a result of an inspection made by the engineer prior to its expiry.

Cost of execution of work of repair etc.

The defective work will be made good at the contractor's expense unless it can be shown that the defect occurred for reasons outside the contractor's control, e.g. design defects. In these circumstances, the work will be valued and paid for on the same basis as other additional work.

Remedy on contractor's failure to carry out work required

Where the contractor fails to do this work, the employer can instruct others firms to carry out the work and then recover the costs from the contractor.

Contractor to search (clause 50)

The engineer may instruct the contractor to carry out searches, tests or trials to determine the nature of any fault or defect. Where such defects are due to the contractor's own negligence, the contractor is responsible for the costs of the tests and the eventual repair of the defective work. Where the work is not due to the contractor's negligence in carrying out the works, the costs of the tests and the remedial action are to be borne by the employer.

Frustration (clause 64)

A contract may be frustrated (see Chapter 1) by war or other supervening event. Under these circumstances, the amount payable by the employer to the contractor in respect of work executed is the same as that calculated under the war clause (clause 65).

CHAPTER 17

COSTS OF CONSTRUCTION

Introduction

An important factor to consider with regard to any major capital works project is the costs involved. In the context of civil engineering contracts, this is usually interpreted to mean the costs of constructing the works in accordance with the contract documents. However, it should be remembered that these represent only a proportion of the costs involved. In addition, the promoter will have to pay the sums of money associated with the purchase of the site. These sums can be considerable, particularly where the planning application for the proposed works is contested, goes to appeal and results in a planning enquiry. Often in the case of civil engineering works, the site for development is determined by several other factors, which result in no other suitable site being available. The promoter will also have to pay the fees of the designers, which may in addition to civil engineers include architects, quantity surveyors, planners, geologists, etc. These can easily add 15–20 per cent to the contractor's tender costs for the project.

Once the project is completed and handed over to the promoter, the promoter may be involved in installation of plant and equipment that were not part of the construction project. Most construction projects include these additional items to a greater or lesser extent. Promoters are now much more aware of the longer-term costs involved in owning and managing a construction project once it becomes their entire responsibility. The proper consideration of the costs-in-use are likely to provide added benefits for a promoter in the longer term.

Budgets

Before and during the design of a civil engineering project it is necessary to estimate the costs involved. There are a wide variety of methods available for this purpose. Most have the advantage of simple quantification that relies upon the expert judgement of those responsible for preparing the early price estimates or budgets.

The estimate or forecast of cost of construction is done at different stages of the project. A budget sum is prepared for the promoter at the inception on behalf of the design team, or design and construct contractor. If this is acceptable more detailed methods are then used to provide a framework for cost planning the entire scheme. This framework may be time-analysed by allocating the costs involved to the various stages of the project. Eventually, the contractor estimates the costs in order to calculate the tender sum. If the tender is successful then this sum is incorporated in the contract documents as the contract sum.

In most cases during the early part of the design stage, the drawings and specifications are uncertain and imprecise. As the design progresses the forecast can be refined, but on average it will vary by at least 10 per cent from the eventual contract sum, which, in itself, is still only an estimate of cost. It is usual to offer a range of possible estimates or confidence limits rather than a single sum. Forecasting, in general, is an imprecise occupation and the forecasting of construction costs is no different. The future is always difficult to forecast and errors and inconsistencies in pricing will occur.

Some of the differences in cost estimates can be accounted for by changes in design and specification, changes in the client's requirements, the introduction of new technologies and inflationary factors. The typical levels of cost estimating accuracy that are achieved in practice are shown in Table 17.1. These can be applied to the full range of different types of construction projects.

Table 17.1 Estimate classification and accuracy

Estimate	Purpose	Accuracy
Order of magnitude	Feasibility studies	+/– 25–40%
Factor estimate	Early stage assessment	+/– 15–25%
Office estimate	Preliminary budget	+/– 10–15%
Definitive estimate	Final budget	+/– 5–10%
Final estimate	Prior to tender	+/– 5%

Contractors' estimating

Contractors' estimating is based upon measuring a large number of work items and analysing their unit costs, based upon previously recorded site performance data. The measured items should include only those which are cost important. In theory the site performance data or labour outputs and material and plant constants are derived from work done on site. Research has shown that this theoretical concept is flawed in practice due to the poor and inappropriate recording systems used by contractors and the lack of confidence that estimators have in individual site feedback. The time taken to undertake the different construction operations is also highly variable. The difficulty of capturing this data in a meaningful form that can hopefully be reused is a complex task beyond the profitable occupation of most contractors. A comparison of similar items priced by different contractors reveals differences or discrepancies of as much as 200 per cent. Even published data on guide prices can vary by as much as 50 per cent.

Tender price

The tender total is defined in clause 1 of the ICE Conditions of Contract. This means the total of the bill of quantities at the date of the award of the contract or, in the absence of a bill of quantities, the agreed estimated total value of the works at that date. The tender total is subsequently adjusted within the terms of the contract to allow for variations described in clauses 51 and 52 (Alterations, additions and omissions). The whole process is aimed at ascertaining the contract price for the works, paid in accordance with the provisions of the contract.

Clauses which are considered in this chapter include those concerned with

- alterations, additions and omissions (clauses 51 and 52)
- measurement (clauses 55–57)
- provisional prime cost sums and Nominated subcontracts (clauses 58–59)
- certificates and payments (clauses 60–61)

ALTERATIONS, ADDITIONS AND OMISSIONS (CLAUSES 51–52)

The alterations, additions and omissions allowed for within the contract are described in the following clauses:

Ordered variations (clause 51)

Most forms and conditions of contract used in the construction industry allow for variations or changes to the design arising at some stage during the construction of the project. The absence of such a clause would necessitate a new contract being drawn up if changes had to be made or did arise during the progress of the works. The disadvantage of such a clause is that it allows the engineer or designer to delay making some decisions until almost the last possible moment. The promoter or owner will be bound by any variations given by the engineer as long as the engineer does not exceed the powers under the terms of the conditions of contract.

The engineer may order variations to any part of the works that are considered necessary for the completion of the works or for any other reasons that may be desirable to improve the functioning of the works. These variations may include:

- additions
- omissions
- substitutions
- alterations
- changes in:
  quality
  form
  character
  kind
  position
  dimension
  level
  line
- changes in:
  specified sequence
  method of timing

Such changes may also be ordered during the defects correction period.

Ordered variations to be in writing

All variations ordered by the engineer must be in writing. Any instruction given by the engineer or by any person exercising delegated duties and authorities under clause 2.4 should be in writing. Oral instructions can be given to the contractor but must be confirmed as soon as possible under the circumstances (clause 2 (6)(b)). The contractor may also confirm such oral instructions in writing. These will be accepted if they are not subsequently contradicted in writing by the engineer (or representative). Under these circumstances the contractor's confirmation of an oral instruction will be deemed to be an instruction in writing by the engineer (clause 2 (6)(b).

Variations not to affect contract

Since this clause is included in the contract, changes to the design or specification, etc. will not invalidate the contract. Only variations ordered by the engineer will affect the contract price. Additional work that is required because of the contractor's own default must be carried out at the contractor's expense.

Change in quantities

Since most civil engineering contracts are of a remeasurement type, changes in the quantities of work described in the bill of quantities do not require a written instruction.

Work that has been described in the contract bills as the main contractor's (or one of the domestic subcontractors should the intention have been to sublet it) cannot be omitted and then awarded to a nominated subcontractor. It can be omitted entirely and done after the contract has been completed. It can also be awarded to a firm employed directly by the promoter, assuming that the main contractor will allow access to the works, for this purpose, during the contract period. Of course, if the main contractor agrees to such a process, this will override the condition.

Valuation of variations (clause 52)

The rules for the valuation of ordered instructions are described in clause 52. The respective values are to be ascertained by the engineer after consultation with the contractor. In practice, the work is measured or re-measured, the contractor is invited to price the work and the final costs are agreed between the two parties. Where the two parties cannot agree, the engineer fixes the rate.

1. Where the additional or substituted work is the same in character, conditions and quantity to items in the contract bills, then bill rates or prices are to be used to value the variation.
2. Where the additional or substituted work is the same in character, but is executed under different conditions or results in significant changes in quantity, then the bill rates or prices are to be used as a basis for valuing the variation. These are known as pro-rata rates.
3. If the additional or substituted work is different from those items in bills, then a fair method of valuation is to be used, often known as star rates.
4. Omissions are valued at bill rates unless the remaining quantities are substantially changed, in which case a revaluation of these items will become necessary using method 3 above.
5. If the varied work cannot be properly measured or valued, the contractor or subcontractor is to be paid for the work on a daywork basis.
6. If the contractor indicates that the rates calculated do not properly reflect the nature of the work due to changes made by variations then the contract allows for the contractor to issue a claim for an additional payment (clause 52 (4)). The contractor will be deemed, in pricing the work at tender stage, to have made due allowance in pricing for the preliminary items. If the execution of any of this work is subject to variation then some adjustment may need to be made to the value of the preliminary items.

The six scenarios described above are shown diagramatically in Figure 17.1.

If any percentage or lump-sum adjustments have been made – for example, in the contract bills – for the correction of errors (see clause 55 (2)) these will need to be adjusted accordingly.

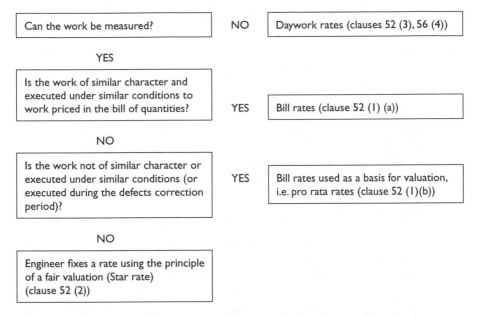

| Can the work be measured? | NO | Daywork rates (clauses 52 (3), 56 (4)) |

YES

| Is the work of similar character and executed under similar conditions to work priced in the bill of quantities? | YES | Bill rates (clause 52 (1) (a)) |

NO

| Is the work not of similar character or executed under similar conditions (or executed during the defects correction period)? | YES | Bill rates used as a basis for valuation, i.e. pro rata rates (clause 52 (1)(b)) |

NO

| Engineer fixes a rate using the principle of a fair valuation (Star rate) (clause 52 (2)) |

Note 1: Work that is omitted from a contract due to a variation will usually be omitted at the bill rates unless the variation changes the nature or quantity of the work that remains (clauses 52 (2), 56 (2)).

Note 2: If the contractor intends to claim a higher rate or price than that calculated using the above principles, the engineer should normally be informed of this within 28 days after giving such notification.

Figure 17.1 Valuation of variations

MEASUREMENT (CLAUSES 55–57)

Quantities (clause 55)

This clause states that the quantities set out in the bill of quantities are only estimated. They thus give the contractor a good guide to the extent of the contract works, but they must not be assumed to be actual or correct quantities of the works.

Correction of errors

Any errors in description or the omission of items from the bill of quantities will not vitiate the contract nor release the contractor from executing the works. Any errors or omissions will be corrected by the engineer in accordance with clause 52 above. These errors or omissions apply only to those made by the engineer or his representative. Any mistakes made by the contractor during the estimating and tendering process remain the responsibility of the contractor and are not corrected.

Measurement and valuation (clause 56)

The responsibility for admeasurement and valuation of the work done in accordance with the contract rests with the engineer.

Increase or decrease of rate

Where the quantities executed in respect of any item in the bill of quantities increase or decrease to such an extent that they are unreasonable to use, the engineer will fix new rates. This is normally done in consultation with the contractor.

Attending for measurement

In practice, remeasurement is frequently done either jointly by the engineer's representative and the contractor's representative or one party measures the work and the other checks it at a later date. It is the engineer's responsibility to determine the value in accordance with the contract. However, contractors who wish to be paid as quickly as possible for the work that they have carried out will often measure the works and submit the information for checking by the engineer. Where the engineer's representative wishes to remeasure work, the contractor's representative will be invited to attend. If this representative fails to attend, the engineer's measurements will be taken to be the correct measurement of the work.

Daywork

Where work is carried out on a daywork basis, the contractor will be paid for the work on the basis of a daywork schedule included within the contract. The 'Schedule of Dayworks carried out incidental to Contract Work' issued by The Federation of Civil Engineering Contractors (FCEC) is usually adopted. Provision is made for contractors to allow for overheads and profit on daywork accounts within the tender documents.

This schedule is more comprehensive than that issued by the Royal Institution of Chartered Surveyors (RICS) and the former Building Employers' Confederation (BEC) for use on building contracts. There are also differences in the way that overheads and profit are to be calculated and included within a daywork account. Since the FCEC and BEC have recently merged to form the Construction Employers' Confederation (CEC) it is not yet known whether the two separate definitions of daywork will continue to be used.

In some cases because of the nature of the work involved and the appropriate engineer's instruction the work may appear to be expensive. The contractor is therefore often requested to submit quotations to the engineer before placing an order for the materials required.

In order for the work to be properly checked and authorised as being satisfactorily completed, the contractor should inform the engineer when work to be carried out on a daywork is envisaged. Upon completion of a daywork the contractor must provide the engineer with records, receipts and such other documentation as may be necessary. This information should be supplied to the engineer within a reasonable time of the work being executed. Vouchers showing the daily time spent on the work, the workmen's names and the plant and materials employed should be verified. It is recommended that this should be done weekly by the engineer's representative on site.

Whilst a contractor can supply a verified voucher, this is for record purposes only. It does not mean that an method of valuation adopted will be that of daywork. This is a matter for an engineer's representative – normally a quantity surveyor – to determine. The use of dayworks may not be the appropriate method by which to value a particular engineer's instruction. Quantity surveyors often contend that work executed on a daywork basis is more costly than similar work undertaken using bill rates, principally because there is no incentive for the contractor to execute the work efficiently and effectively. However, work that is properly classified as daywork is often complicated and time-consuming to perform. Due to its nature it often takes longer to complete and will be thus more expensive.

Method of measurement (clause 57)

The method of measurement used for the preparation of the bills of quantities is assumed to be the Civil Engineering Standard Method of Measurement (second edition, 1985). The CESMM is approved by the Institution of Civil Engineers and the Federation of Civil Engineering Contractors in association with the Association of Consulting Engineers. The relevant SMM and its date of issue is to be stated on the Form of Tender.

There are a large number of methods of measurement that are used for measuring construction work in the United Kingdom. These include: the Standard Method of Measurement of Building Works, known colloquially as SMM7 (1988), the Standard Method of Measurement of Industrial Engineering Construction (1984) and the Manual of Contract Documents for Highway Works (1991). This last method is the second most important SMM for works of civil engineering construction. Many overseas countries have developed their own methods of measurement, often using UK methods as a basis.

PROVISIONAL AND PRIME COSTS SUMS AND NOMINATED SUB-CONTRACTS (CLAUSES 58–59)

Use of provisional and prime cost sums (clause 58)

The engineer may issue instructions that a provisional sum inserted into the bill of quantities be executed by the main contractor or by a nominated subcontractor.

Table 17.2 Provisional and prime costs sums

	Provisional sums	Prime cost sums
Clause reference	Clause 58 (1)	Clause 58 (2)
Definition clause	Clause 1 (l)	Clause 1 (k)
Characteristics	Contingency sum used at the discretion of the engineer	Sum available for the execution of work normally of a specialist nature
Carried out by	Main contractor	Specialist subcontractor
Basis of payment	Bill rates (see Figure 17.1)	Quotation and invoice

Use of prime cost sums

Prime cost sums, usually intended for highly specialised work, can be dealt with in a similar way. In the case of provisional sums the work is normally paid for on the basis of the rates and prices inserted in the bill of quantities, although a contractor could be asked to submit a quotation for the work before it is carried out. In the case of prime cost sums, the contractor would normally be expected to submit a quotation and to be paid for the work on this basis. In this way the contractor's quotation would be easily compared with that of the other potential nominated subcontractors.

Design requirements to be expressly stated

The expenditure of any prime cost or provisional sum may also include design and specification work on the part of the contractor or nominated subcontractor, in addition to the more usual construction work. The obligation of the contractor is restricted to that expressly stated in this clause.

Nominated subcontractors (clause 59)

Nominated subcontractors are defined in clause (1m) and include firms which execute work or which supply goods, materials or services for which a prime cost has been included in the contract. Building contracts separate firms which are nominated to supply goods and materials (nominated supplier) from those which are nominated to supply goods and materials and execute work (nominated subcontractor). The ICE form of contract makes no distinction between suppliers and subcontractors and classifies all such firms as nominated subcontractors.

Main contractors are also unlikely to complete all of the work described in the bills of quantities using their own resources. Some firms will therefore be appointed direct by a contractor to carry out aspects of the project. These firms are described as domestic subcontractors (see clause 4).

Contractor's objections to a nominated subcontractor

Contractors may have a reasonable objection against nominating a particular firm as a subcontractor. If the objection is upheld, the engineer must appoint a different firm as a subcontractor. If a subcontractor refuses to enter a sub-contract containing the following provisions, the engineer must also appoint a different firm, which is willing to enter into such an agreement.

- Obligations and liabilities are similar to those of the main contract.
- Contractor to be indemnified against all claims, demands and proceedings and the costs and expenses involved.
- There is to be no negligence on the part of the subcontractor or misuse of the contractor's equipment or temporary works.
- Security provided for the proper performance of the sub-contract.
- Provisions equivalent to those contained in clause 63 regarding the determination of the contractor's employment.

Engineer's action upon objection to nomination or upon determination of nominated subcontract

Where a main contractor objects to the appointment of a particular subcontractor the engineer may:

- nominate an alternative subcontractor
- vary the works in accordance with clause 51
- omit the work from the contract and carry out the work as a separate contract with the employer
  - this may be done concurrently with the main contract, in which case clause 31 will apply, or the work may be executed at some later date
  - the contractor may be reimbursed for a loss of profit
- instruct the main contractor to appoint a subcontractor
- invite the contractor to execute the work

Contractor responsible for nominated subcontract

The main contractor is eventually responsible for the work of all subcontractors other than that outlined in clause 58 (3) which covers design and specification work.

Nominated subcontractor's default

These clauses outline the procedures to be followed in the event of a default made by a subcontractor. The contractor should first notify the engineer in writing and the engineer's consent is required before a subcontract can be terminated.

Termination of subcontract

The engineer may expel the subcontractor from the project and rescind the subcontract. If the engineer's consent is withheld, the contractor will require instructions under clause 13.

Engineer's action upon termination

Where a subcontract is terminated, the engineer should immediately take the appropriate action required under clause 59 (2) and described above.

Recovery of additional expense

Having received the engineer's consent for the termination of the nominated subcontract, the contractor must take all available steps to recover the additional expenses incurred under the subcontract. These may include the employer's expenses resulting from the termination.

Reimbursement of contractor's loss

If the contractor fails to recover the reasonable expenses involved in completing the subcontract works, such as the appointment of other firms to carry out the work, as well as the costs involved in the termination, these will be reimbursed by the employer.

Consequent delay

The engineer will take such a termination into account when determining any extension of time to which the contractor might be entitled under clause 44.

Provisions for payment

These costs might include the following items:

- the actual price paid by the contractor to the subcontractor for the work properly executed within the terms of the contract, based upon an invoice for the work
- a cash discount payable to the contractor for prompt payment; all other trade discounts, rebates and allowances are excluded
- any charges itemised in the bill of quantities for the fixing of materials in the case of a nominated subcontractor who was appointed only to supply goods or materials
- a profit percentage where a rate has been set against the item in the bill of quantities; this is also listed in Appendix 2 to the form of contract

Production of vouchers

The contractor must provide the engineer when required with quotations, invoices, vouchers, sub-contract documents, accounts and receipts for all nominated subcontracts.

Payment to nominated subcontractors

Nominated subcontractors are paid through the main contract at the appropriate interim payment stage (clause 60). Before issuing any certificate the engineer should check that amounts included in previous interim certificates for nominated subcontractors have been paid by the contractor. Where a contractor cannot satisfy the engineer in this respect or supply information to justify a refusal to make such a payment, the employer can pay the nominated subcontractor direct. Such amounts are then deducted by way of set-off from amounts owed by the employer to the contractor under subsequent certificates.

CERTIFICATES AND PAYMENTS (CLAUSES 60–61)

Monthly statements (clause 60)

The contractor shall submit at monthly intervals a statement showing:

- the estimated contract value of the permanent works executed up to the end of that month
- a list of any goods or materials delivered to the site but not yet incorporated into the permanent works. These are limited by a percentage to be stated in the Appendix part 2 (item 13) to the form of tender.
- a list of goods and materials not yet delivered to the site but of which the property has been vested in the employer; these are identified in the Appendix to the form of tender
- the estimated amounts of other items in the contract such as temporary works or contractor's equipment for which separate amounts have been included in the bills of quantities
- amounts in respect of nominated subcontracts are to be shown separately in the certificate

Monthly payments

Interim valuations of the work completed by the contractor are generally prepared by the contractor's quantity surveyor or measurement engineer. These are then agreed by the engineer or the engineer's representative, usually a quantity surveyor.

Minimum amount of certificate

Whilst the frequency of interim certificates is usually monthly, the contractor may decide that insufficient work has been carried out and a certificate is not justified. The conditions of contract also stipulate that the valuation must achieve a minimum amount, which is to be stated on the form of tender Appendix 1, item 14, before any monthly payment will be made. The purpose of such a recommendation is to attempt to ensure that the contractor is making regular progress with the works. On very large contracts it may be necessary to provide approximate valuations on a weekly basis in order to satisfy the contractor's cash flow. More accurate payments would continue to be made at the monthly interval. This arrangement may be required at peak construction periods, for example during the summer months on mass earth-moving contracts. Whilst the Conditions of Contract are silent on this matter, in practice this is a common occurrence for very large projects.

Final account

It is the contractor's responsibility to prepare the final account and to provide the supporting documentation. This will show in detail the value of the work executed under the contract. The account should be provided within 3 months of the defects correction certificate. Within a further 3 months the engineer will issue a certificate stating the final amount due to the contractor from the employer. This amount should be paid within 28 days of the date of the certificate, less any amounts for liquidated damages for delay as described in clause 47. It will be noted above that half the retention money owing at the end of the defects correction period will have been paid within 14 days of the issue of the defects correction certificate.

Where the work has been accurately valued at the issue of the certificate of substantial completion, only retention monies will be outstanding and these can be released with the final certificate. There will therefore be no requirements to issue a certificate for further payments between these two certificates. The certificate of substantial completion has often been referred to as the penultimate certificate, although there is nothing in the contract to prohibit the issue of further certificates beyond this one. In a real contract situation it is often not possible to value accurately the works at substantial completion, and so further certificates are issued.

Retention

The contractor is not paid the full amount of the monthly statement, but this is subject to a deduction in the form of retention. Retention is applied to work that has not reached practical completion and to materials and goods referred to above, including nominated works.

The purpose of this retention is:

- to provide some incentive for the contractor to complete the works
- to provide some sort of security should the contractor default

The employer has only a fiduciary interest in the retention. In this respect the retention is held on trust only, and need not therefore be invested to accrue interest. The amount retained by the employer under this clause is to be stated on the form of tender Appendix (Part 1) items 15 and 16. It is recommended that it should not exceed 5 per cent with a limit of 3 per cent of the tender total. The argument used for this arrangement is that once a retention fund has been established this is sufficient to meet the two purposes listed above.

> Example: Tender sum £3,000,000. 5 per cent retention, 3 per cent retention fund limit. 3 per cent of £3m is £90,000. When the work completed reaches £1,800,000 then 5 per cent retention will equal the retention fund limit. Work completed after this will then be paid in full to the contractor without any further deduction of retention.

A different, higher or lower, rate of retention can be agreed between the parties. The figures described above represent the percentages recommended in the contract.

Payment of retention

Retention is released to the contractor in the following ways:

- Upon the issue of a certificate of substantial completion in respect of any section or part of the works, one half of the relevant retention is to be paid.
- Upon the issue of the certificate of substantial completion in respect of the whole of the works one half of the total retention is to be paid within 14 days of the issue of this certificate.
- At the expiry of the defects correction period the remainder of the retention fund is to be paid to the contractor within 14 days.
- Retention is released to the contractor even where claims by the contractor are still outstanding.

Payment

The contractor is to be paid the amount of the agreed interim certificate within 28 days of submitting a monthly statement to the engineer. The amount paid to the contractor is calculated as follows:

- the agreed contractor's monthly statement, described above
- less retention sums, described above
- less any payments previously made to the contractor

Where the amount paid to the contractor differs from the amount certified by the engineer, the employer must provide written information as to how the sum has been calculated.

A copy of the certificate, issued by the engineer, is sent to the employer and the contractor, with a detailed explanation where necessary.

In some contract forms, such as the Joint Contracts Tribunal form (JCT80) there is provision for payment at defined stages of completion of the project. This may be used in preference to valuations which are prepared on a monthly basis or on whatever basis has been stated in the contract. There is no similar provision in the ICE form.

Interest on overdue payments

Where either the engineer or the employer fails to make payment to the contractor, then the employer is to pay the contractor for the loss of interest. This is calculated at 2 per cent above the base lending rate specified in the Appendix to the form of tender, item 17. The interest due is calculated on a compound interest basis. If an arbitration pursuant to clause 66 recommends a sum due that should have been previously certified, interest should be added on the same basis.

Correction and withholding of certificates

The engineer must not include in any certificate work that does not comply with the contract or work that is of an unsatisfactory nature. Quantity surveyors or measurement engineers who may be responsible for preparing and agreeing contractor's interim statements of work should not knowingly include work that is unsatisfactory. The engineer's assistant has no responsibility to withhold payment in this way but would be well advised to bring the matter to the attention of the engineer.

However, once work has been included within a certificate for a nominated subcontractor it cannot normally be reduced in subsequent certificates, if the contractor has already paid the amount to the nominated subcontractor. If in the final certificate the engineer does reduce such sums, then the employer must reimburse the contractor by the same amount together with interest at the agreed rate.

Copy of certificate for contractor

The contractor and the employer are to receive a copy of every certificate that is issued by the engineer.

Payment advice

The employer must notify the contractor immediately of any differences between the payment made to the contractor and the amount certified by the engineer.

Defects correction certificate (clause 61)

The defects correction certificate is issued either at the end of the defects correction period or once all the defects have all been made good, whichever date is the later.

When issuing the defects correction certificate the engineer will state the date on which the contractor's obligations to construct and complete the works to the engineer's satisfaction were completed.

Unfulfilled obligations

The issue of this certificate does not relieve either the contractor or the employer from any liability towards each other under the terms of the contract.

Contract price fluctuations

The ICE Conditions provide for separate clauses where it is necessary to make special provision to take account of price fluctuations.

PART FIVE

OTHER CONDITIONS OF CONTRACT

CHAPTER 18

GENERAL CONDITIONS OF CONTRACT FOR BUILDING AND CIVIL ENGINEERING

Introduction

Public works projects represent a significant sector of the construction industry and a large proportion of civil engineering contracts. The various types of projects are undertaken by a variety of organisations both locally and centrally. Local authorities have a preference for ICE Conditions of Contract for civil engineering works and JCT 80 (Local Authorities Edition) for building works. Construction projects administered by central government agencies, however, prefer to use their own forms of contract, i.e. General Conditions of Contract for Building and Civil Engineering Works. The latest edition was published in 1990 and revised in 1991. The forms which are currently in use are:

- GC/Works/1/ – third edition which is for use on major projects. There are different versions, depending upon whether the quantities are firm, subject to remeasurement or without quantities, and single-stage design and build
- GC/Works/2/ – for use on minor works

A further form is in use for the measured-term type contracts, which are an important feature of maintenance projects carried out for government establishments within a specified contract period. Central government pioneered this type of contractual arrangement, which is now used by other types of employers.

The forms are now prepared under the auspices of the Department of the Environment and published by Her Majesty's Stationery Office. The Conditions do not include a form of tender or agreement but these are provided for separately. This latest edition of the Conditions envisages the use of a PM or project manager replacing the term engineer under the ICE Conditions of Contract. The PM is defined as being the manager and superintendent of the works. If the works have been fully designed (the implication in most of the forms in use!) then there is no necessity for the PM to be restricted to a designer. The appointment to this key position during the construction phase can then be given to the individual who is most suited to such a role. The names of the various parties associated with the project are given in the Abstract of Particulars.

The clauses, referred to as conditions, are numbered consecutively from 1 to 65 and are grouped together under eight major sections. Conditions, such as a variation of price condition, may be added to supplement the printed General Conditions according to circumstances. They are incorporated into the contract conditions by listing in the Abstract of Particulars, which is included with the invitation to tender. The clauses contain the usual amalgam of items that are to be found in the other forms. However, it is generally accepted that they are more legally precise than the other forms. Some of the conditions are similar in their effect to ICE Conditions, e.g. setting out, patents, antiquities. Other clauses have no parallel in the ICE Conditions. Some of these cover matters of a procedural nature and include, for example, passes, which are now a standard security measure of all government agencies, racial discrimination, photographs and emergency powers. Some clauses included in ICE Conditions have no equivalent in this form, for example, the war damage clause. This matter, should it occur, is dealt with independently by the government department concerned. The following is a brief résumé of some of these contract conditions, from GC/Works/1.

Contract, documentation, information and staff

Condition 1 lists a number of definitions which are referred to throughout the form of contract. Many other forms of contract now adopt this approach in order to clarify matters which are relevant to the execution of the project.

Conditions 2 and 3 are concerned with the contract documents and their relationship with each other. The documents are listed in the definitions under 'contract' and include:

- contract conditions
- Abstract of Particulars
- specification
- drawings
- bills of quantities
- the tender
- authority's written acceptance

In common with the ICE Conditions both a specification and bills of quantities are contract documents. The conditions of the contract are the most important, followed by the specification and then the drawings. The PM can reverse the importance of these last two documents should this be desired.

Condition 3 refers to bills of quantities and deals with the method of measurement (which needs to be identified). It also describes how errors and omissions should be dealt with and the use of approximate quantities.

Conditions 4–6 deal with staff matters and include the delegation and representation of the authority's powers, the requirement of the contractor to keep a competent agent on site to supervise the execution of the works, and the removal from site of anyone whom the PM thinks is undesirable. On this last matter, there is little scope for negotiation since the PM's view is final and conclusive.

General obligations

Condition 7 requires the contractor to be fully satisfied on the following points regarding the site:

- communications and access to it
- its contours and boundaries
- the risks associated with adjacent property and its occupiers
- the ground conditions
- conditions under which the work will be executed and the precautions necessary to prevent nuisance and pollution
- availability of labour
- availability of goods, materials, etc.
- any other factors which will influence the execution of the works or the tender price

No claim for additional payments will be allowed if the contractor misunderstands or misinterprets any of the above. However, if the contractor encounters unforeseeable ground conditions during the execution of the works, the PM may agree to the contract sum being increased or decreased accordingly.

Condition 8 covers matters of insurance and specifically:

- employer's liability insurance
- insurance against loss or damage
- insurance against personal injury and loss or damage to property

Condition 9 outlines the various responsibilities in respect of setting out. The PM provides the information for setting out the works. It is the contractor's responsibility to do the setting out, including providing all the necessary instruments, profiles, templates, etc.

Condition 10 allows for some of the design to be done by the contractor or subcontractor, should this be required. The contractor's liability in this respect is the same as if the design work had been provided by an architect.

Conditions 11–12 describe the need to comply with statutory provisions and pay the charges involved and to make the necessary payments for royalties or patent rights.

Conditions 13–14 require the contractor to protect the works and goods and material on site from damage, and to protect workpeople and the public from any danger. The conditions also require the contractor to prevent nuisances and inconvenience to anyone.

Conditions 15–17 require the contractor to inform the PM of the number and type of workpeople employed on the site each day, that the excavations are ready to receive the foundations and that work which should be inspected by the PM is ready to be covered up with earth or other materials.

Conditions 18 and 25 cover the meeting of the quantity surveyor or measurement engineer and the contractor's representative to agree measurements on site, and the keeping of contract records which may be needed for the preparation of the final account. In the absence of the contractor's representative, the measurements can still be made and used as if these had been agreed.

Condition 19 is concerned with:

- loss of and damage to property
- personal injury to, or sickness or death of, any person
- loss of and damage to the works, including materials on site
- loss of profits or loss suffered because of any loss or damage

Contractors must take steps at their own expense to reinstate, replace or make good to the satisfaction of the authority any loss or damage to the works. If this also results in a claim against the authority by third persons, the contractor has to reimburse the authority for any costs or expense that may be incurred. The contractor can be reimbursed if the loss or damage was the fault of the authority, or due to unforeseen ground conditions, an accepted risk or circumstances outside the control of the contractor.

Conditions 20, 22 and 23 are conditions not found in other forms of contract. Condition 20 outlines the contractor's position in respect of the loss, destruction or disclosure of 'personal data' and access to it, as defined in the Data Protection Act 1984. Condition 22 deals with contractors working within the boundaries of existing government premises and condition 23 covers the scope of the Race Relations Act 1976.

Condition 21 deals with the making good of defects during the maintenance period, the contractor's responsibility for the costs involved and in the case of his default, the recovering of costs by the authority.

Condition 24 deals with corruption and the receiving of gifts, considerations, inducements or rewards for doing nothing or for showing favour. The authority need be only reasonably satisfied that a breach has occurred, rather than needing to rely on the provisions of the Prevention of Corruption Acts 1889 and 1916 in order to determine the contractor's employment.

Security

Conditions 26–29 cover matters of a government security nature, which are naturally not to be found in the other forms of contract. These include unauthorised admittance to the site, the need for passes in some site locations, limitations on photographs of the site and a general awareness of the Official Secrets Act 1911 and 1939.

Materials and workmanship

Condition 30 deals with the vesting of 'things' which have been defined in condition 1 of the form of contract. 'Things' has two different meanings:

- 'Things for incorporation' means goods and materials intended to form part of the completed work.
- 'Things not for incorporation' means good or materials provided or used to facilitate execution of the works but not for incorporation in them.

The contractor is responsible for the protection, preservation, and replacement of things that are lost, stolen, damaged, destroyed or unfit or unsuitable for their intended purpose. Nothing must removed from the site without the written consent of the PM.

Condition 31 is the main condition which deals with quality. It states that the contractor shall execute the works:

- with diligence
- in accordance with the programme
- with all proper skill and care
- in a workmanlike manner

The contractor must use the necessary skill and care to ensure that the works and anything that is due for incorporation conforms to the requirements of the specification, bills of quantities and drawings. The PM has the power at any time to inspect, examine or test on site at a factory or in a workshop. The PM can arrange for an independent expert to do the testing in order to ensure conformity with the contract requirements. In the event of a failure, the contractor has to bear the costs of the test and any retesting that may ensue.

Condition 32 refers to excavations, which in this context means antiquities such as fossils and other items of interest or value found on the site. Such findings remain the property of the employer.

Commencement, programme, delays and completion

Conditions 33 and 35 deal with the programme and progress of the works. The contractor must submit a programme to show the sequence of operations, details of temporary work, method of working, labour and plant requirements and the critical events which might influence the progress and completion of the works. Progress meetings are normally held once a month where the contractor is required to provide a written report showing any or all of the following:

- the relationship between progress of the works and the contract programme
- information required by the contractor
- delays and possible delays

- requests for an extension of time
- proposals to bring the project back on schedule

The PM must provide a written statement in response to these points within 7 days of this meeting.

Conditions 34 and 36–38 describe commencement, completion, extensions of time, early possession and acceleration. The authority will notify the contractor of when possession can be effected and completion is then calculated from this time in accordance with the contract. Condition 36 identifies the reasons for granting an extension of time to the contractor. This is only done after first receiving a request for such from the contractor. Early possession of the project is covered in condition 37; where a section or a part of the works is completed to the PM's satisfaction and the authority wishes to take possession it is referred to as a completed part. This has the effect of starting the maintenance period for that part of the works, restricting any further liquidated damages and releasing part of the reserve, i.e. one-half of the retention money which is held on that part of the project. Condition 38 allows for achieving an accelerated completion date. If this is required, the contractor is asked to price specified proposals and indicate how early completion might be achieved.

Condition 39 covers matters relating to the issue of certificates at completion and at the end of the maintenance period.

Instructions and payments

Condition 40 lists the procedures dealing with the PM's instructions. The provisions allow for the issue of further drawings, details, instructions, directions and explanations. Instructions must be in writing or confirmed in writing within 7 days. Instructions which alter, add, omit or change the design, quality or quantity of works are known as variation instructions (VIs). Instructions may be given in respect of the following:

- the variation/modification of the specification drawings, bills of quantities, or the design, quality or quantity of works
- discrepancy between specification, drawings and bills of quantities
- removal from site of any things for incorporation and their substitution with any other things
- removal/re-execution of work
- order of execution of the works
- hours of work and overtime
- suspension of the works
- replacement of any person employed in connection with the contract
- opening up for inspection
- making good of defects
- execution of emergency work
- use/disposal of materials from the excavations

- action to be followed on the discovery of fossils, etc.
- measures to avoid nuisance/pollution
- any other matter which the PM considers is expedient

Conditions 41–43 cover matters relating to the valuation of work. The valuation of a VI can be determined in one of two ways:

- the acceptance by the PM of a lump-sum quotation
- valuation by the measurement engineer or quantity surveyor using the following principles:
  - use of bill rates
  - use of rates which are pro-rata to rates in the bills
  - fair valuation or agreement
  - daywork rates
  - adjustment to these rates for any disruption

If circumstances arise where, as a result of an instruction which is not a VI, the contractor properly incurs expense beyond that provided for, or makes a saving in the cost of executing the works, such costs can also be adjusted.

Condition 44 covers labour tax, i.e. any tax, levy or contribution which by law has to be paid by the contractor. If these vary during the course of the contract, the appropriate sum is added to or deducted from the contract sum.

Condition 45 consists of the usual conditions describing VAT.

Condition 46 covers prolongation and disruption to the works and the associated costs.

Condition 47 deals with the honouring of payments due to the contractor. If the authority withholds payments, the contractor is allowed to add finance charges to those amounts at a rate of 1 per cent above the Bank of England's lending rate. Condition 48 deals with interim monthly payments as follows:

- 95 per cent of the relevant amount from a stage payment chart, assuming that the project is on schedule. The remaining 5 per cent is defined as a reserve; 100 per cent of variations
- 100 per cent of prolongation and disruption
- 100 per cent of finance charges

Condition 49 covers the preparation of the final account. Upon completion of the works half of the reserve fund is paid to the contractor. The final account should normally be prepared within 6 months from this date, and any difference between this and the amount paid at completion will be paid to the contractor as soon as possible. The remainder of the reserve fund is paid at the end of the maintenance period.

Conditions 50–52 cover the issue of certificates by the PM and the ability to recover sums owing by the contractor against these certificates, even from other projects. Condition 52 deals with suggestions which the contractor may have for effecting cost savings either in the works or in the future maintenance costs of the project.

Particular powers and remedies

Condition 53 states that, if the main contractor fails within a reasonable period of time to comply with an instruction, the contract may be terminated.

Condition 54 provides for emergency work to be undertaken as required, by the PM.

Condition 55 outlines the procedures to be adopted in the case of a delay in completing the works and the application of liquidated damages. These are included in the Abstract of Particulars, and the appropriate amounts may be deducted from advances paid under interim certificates.

Conditions 56–58 cover determination by the authority (there is no comparable provision for determination by the contractor) and the procedures to be followed in terms of contract completion and financial arrangements. There are circumstances, however, where the contractor may suffer unavoidable loss due to the determination and the contract makes provision for reimbursement where this is thought to be reasonable.

Conditions 59 and 60 describe the procedures to be followed in response to disputes arising between the parties. Initially, matters are referred to the adjudicator who has been named in the Abstract of Particulars. If the matter cannot be satisfactorily resolved at this stage, the dispute can be referred to arbitration.

Assignment, subletting, subcontracting, suppliers and others

Conditions 61–65 deal with those employed on the site other than the main contractor. Condition 61 forbids assigning the contract without authority. Condition 62 covers subletting and the need to obtain authority to do so and to ensure that the main contract conditions are fully covered in each subcontract. Condition 63 describes the procedure to be used in the case of nominated suppliers and subcontractors. These include the usual provisions of the authority paying for the work direct and the reasonable objection that a main contractor may have in entering a contract with a particular firm. The conditions make no references either to any main contractor's attendance or to cash discounts. Condition 64 refers to provisional sums, the need for instructions from the PM prior to their execution and their subsequent valuation under condition 42. Condition 65 gives the authority powers to execute other work on site at the same time as the works are being executed.

CHAPTER 19

ICE CONDITIONS OF CONTRACT FOR MINOR WORKS

Introduction

This form of contract was first introduced in 1988. It is sponsored by the:

- Institution of Civil Engineers (ICE)
- Association of Consulting Engineers (ACE)
- Federation of Civil Engineering Contractors (FCEC)

The use of these conditions of contract is normally limited to minor works of civil engineering construction in the United Kingdom. There is no definition of what constitutes minor works. The term implies smallness in terms of tender costs and the level of complexity of the project. The current guidance notes suggest a contract value not exceeding £250,000 and a contract period not exceeding 6 months. A permanent joint committee of the sponsoring authorities aims to keep the use of the document under review and to consider suggestions for its amendment or revision. The second and current edition of these conditions was introduced in 1995 to provide consistency with changes to the ICE Conditions of Contract used on major civil engineering projects.

The Conditions of Contract for Minor Works are arranged in six parts:

- the Agreement
- the contract schedule
- the thirteen contract conditions
- the Appendix
- the guidance notes
- Conciliation Procedure (1994)

These Conditions of Contract, in common with other ICE forms, include a useful index.

Agreement

This is the formal agreement between the two parties, namely the employer and the contractor. It identifies the project to be constructed and includes three articles of agreement, as follows:

- Article 1: The contractor, subject to the conditions of contract, agrees to perform and complete the works.
- Article 2: The employer will pay the contractor the amounts due under the Conditions of Contract.
- Article 3: The documents listed in the contract schedule form part of the Agreement.

Contract schedule

This lists the documents that are effectively the contract document under this form. These include the:

- Agreement
- contractor's tender. This does not include any separate conditions of the contractor, unless they are agreed in writing to be incorporated in the contract.
- conditions of contract
- Appendix to the conditions of contract
- drawings, including their reference numbers
- specification
- priced bills of quantities *
- schedule of rates *
- daywork schedules *
- Any correspondence, which must be listed *

* Where these are not directly applicable they should be deleted

Clauses

Definitions (clause 1)

Five definitions are provided:

- Works: everything that is necessary for the completion of the contract, including variations
- Contract: the Agreement, conditions of contract, the Appendix and items listed in the contract schedule
- Cost: includes overhead costs, whether on or off site but excluding profit
- Site: places on, under, in or through which the works are to be executed
- Excepted risks: These are:
  - the use or occupation by the employer of any part of the works
  - any defect, error or omission in the design of the works
  - riot, war, invasion, act of foreign enemies, hostilities
  - civil war, rebellion, revolution, insurrection, military or usurped power
  - ionising radiations, contamination from radioactivity
  - pressure waves from aircraft or other aerial devices

Engineer (clause 2)

The employer must appoint an individual to act as the engineer and inform the contractor accordingly. Where a replacement engineer is required for any reason the contractor must be notified in writing of this person. The engineer may appoint a named resident engineer to watch and inspect the works and have delegated powers from the engineer. These powers must be given in writing to the contractor. The engineer has power to issue instructions for:

- variations to the works
- carrying out tests and investigations
- changing the intended sequence of the works
- measures necessary to deal with obstructions (clause 3.8)
- removal of materials that are not in accordance with the contract
- removal of works that are not in accordance with the contract
- explanation to enable the contractor to complete the works
- reasonable exclusion from site of any person

The engineer, upon the written request of the contractor, may need to specify under which of the foregoing powers any instruction is given. Where appropriate, the engineer may instruct that some of the work is to be carried out on a daywork basis, in accordance with the daywork schedule included in the contract. This will normally adopt the Schedule of Dayworks carried out incidental to contract work, that is issued by the former FCEC.

The engineer may order the suspension of the works for the following:

- proper execution of the works
- for the safety of the works
- by reason of weather conditions

Instructions may be issued to protect and secure the works under these circumstances. Where the period of suspension lasts for 60 days, the contractor may serve a written notice for permission to proceed. If after a further 14 days and where the contractor is not in default, permission is not granted, then the contractor can assume that the contract has been abandoned by the employer. Each party to the contract must comply with the engineer's instructions unless:

- it is altered or amended by settlement under clause 11
- it is altered by the decision of the arbitrator under clause 11

General obligations (clause 3)

The following are the general obligations of the contractor:

- perform and complete the works
- provide all supervision, labour, materials, plant, transport and temporary works

- take full responsibility for and the care of the works, until 14 days after the issue of the certificate of practical completion of the works
- take full responsibility for and care of any outstanding work which may have to be finished during the defects correction period
- repair and make good defective work at the contractor's expense
- repair and make good defective work, arising from excepted risks, at the employer's expense
- make good any work damaged during the completion of any outstanding work
- inform the engineer of the name of the contractor's representative
- take full responsibility for the setting out of the works
- take full responsibility for the adequacy, stability and safety of site operations and methods of construction
- take responsibility for the design of temporary works
- give written notice of adverse physical conditions
- afford reasonable facilities to other contractors working on the site

The following are the general obligations of the engineer:

- issue the certificate of practical completion of the works
- issue a certificate of practical completion of a part of the works (where appropriate)
- responsible for the issue of instructions, drawings and other information
- responsible for the design of the works (other than permanent or temporary works designed by the contractor)
- pay for the additional costs of encountering adverse physical conditions, where an experienced contractor could not have reasonably foreseen them
- pay the reasonable costs of any delay or disruption to the works in connection with the above

Starting and completion (clause 4)

The starting date is usually given in the Appendix. The contractor should start work on site within a reasonable time after the starting date. The contractor must provide a programme within 14 days of the starting date. The contractor must proceed with the work in accordance with the programme or with any modifications that the engineer may request. The period for completion is stated in the Appendix and may be extended for one of the reasons that are listed below:

- engineer's instruction
- where a test or investigation complies with the contract
- encountering an obstruction or adverse physical conditions
- delay in the receipt of information
- failure on the part of the employer to give access to the works
- delay in receipt of materials provided by the employer

- exceptionally adverse weather
- circumstances outside the control of the contractor

The contractor must take all reasonable steps to minimise any delay. The extended period of time will be kept under review by the engineer.

Practical completion of the works (or a part of the works) occurs when the employer takes possession of the project. Some minor outstanding items of work may still need to completed at this stage. The engineer must promptly certify this in writing to the contractor. Where the amount of outstanding work prohibits the issue of this certification, the engineer should specify what is required to achieve practical completion. Where the works, or part of the works, are not completed on time, the employer may be entitled to liquidated damages. The amount of these are stated in the Appendix. After the issue of the certificate, the contractor must rectify any defects and complete any outstanding work.

Defects (clause 5)

The defects correction period commences at the completion of the project for the length of time stated in the appendix or for however long it takes for the defects to be made good. Where defects appear during this period, due to materials or workmanship that are not in accordance with the contract, they are to be made good at the contractor's expense. The contractor must correct such defects within a reasonable time. Where, after a written notice from the employer, the work is not carried out, other firms can be employed and the costs deducted from amounts owed to the contractor. At the end of the defects correction period and when all the work has been made good, the engineer must issue a certificate stating that the contractor has complied with all of the obligations under the contract. This does not affect the rights of either party in respect of defects appearing after the defects correction period.

Additional payments (clause 6)

The engineer should agree fair and reasonable amounts with the contractor in respect of additional works or additional costs arising from delay or disruption to the progress of the works as a result of the following:

- engineer's instruction
- where a test or investigation complies with the contract
- delay in the receipt of information
- failure on the part of the employer to give access to the works
- delay in receipt of materials provided by the employer

Where any work is omitted, a reasonable deduction should be made from the contract.

Payment (clause 7)

The works are to be valued as provided in the contract. Payments are to be made monthly, using information supplied by the contractor as a basis. This information will include the value of works completed and goods and materials delivered to site. Within 28 days the employer should pay the amount agreed, less retention sums, at the amounts stated in the Appendix. The Appendix provides for a rate of retention and a limit of retention. Payments will only be made where they reach the minimum amount stated in the Appendix. Half of the retention monies is paid on completion of the project and the remainder with the issue of the certificate that is issued at the end of the defects correction period. Within 28 days of the issue of the certificate at the end of the defects correction period, the contractor should submit to the engineer a final account. This should include supporting documentation to enable the engineer to ascertain the final cost of the project. Within 42 days, the engineer should issue the final certificate. The employer must pay this amount within a further 14 days. Except in cases of fraud or dishonesty, the certificate is conclusive evidence of the amount due to the contractor.

Where payments are not made to the contractor at the appropriate times, interest is added at 2 per cent above the amount of the bank base lending rate stated in the Appendix. In addition to payments due under certificates, the employer must separately identify and pay to the contractor any VAT that is properly chargeable by the Commissioners of Customs and Excise.

Assignment and subcontracting (clause 8)

Neither the employer nor the contractor can assign the contract without the written consent of the other party. The contractor must obtain consent from the engineer to sublet parts of the works. The whole of the works cannot be sublet. The contractor remains responsible for all subcontractors' work and where a subcontractor will not rectify defects the contractor must do this at his own expense.

Statutory obligations (clause 9)

The contractor must comply with all notices required by statutory authorities and pay all fees that may be necessary. The contractor is not liable for any failure to comply with statutory requirements where these are carried out in accordance with the contract or on the instruction of the engineer. It is the employer's responsibility to obtain, within due time, any consent, approval, licence or permission that may be required for the execution of the works.

Liabilities and insurance (clause 10)

The contractor is to maintain an insurance in the joint names of the employer and the contractor, for both the permanent and the temporary works.

This insurance will extend throughout the contract period. It must also cover any loss or damage caused by the contractor up to the end of the defects correction period. The contractor must indemnify the employer against all losses and claims for injury or damage to any person or property whatsoever. Where the engineer or employer has contributed towards such a loss the contractor's liability will be reduced accordingly. The contractor is not liable in respect of the following:

- damage to crops on the site
- use or occupation of land for the purpose of constructing the works
- interference with any easement or quasi easement
- right of the employer to construct works on, over, under, in or through any land
- damage which is unavoidable during construction
- injuries or damage resulting from the neglect of the employer or engineer

The employer must also indemnify the contractor against all claims, demands, proceedings, damages, costs, charges or expenses in respect of the above. The contractor's insurance must be with a company that is approved by the employer. The amount of insurance is stated by the employer in the Appendix. The terms of insurance must include provision whereby, if the contractor is entitled to receive indemnity against the employer, then such claims will be met. The contractor must comply with the terms of the policy and provide evidence that premiums have been paid. The policy of insurance can be inspected at any time by the employer.

Disputes (clause 11)

Disputes that may arise between the parties are to dealt with by conciliation or arbitration. Where appropriate, the ICE Conciliation Procedure 1994 or the ICE Arbitration Procedure 1983 should be used. A dispute is deemed to arise where one party to the contract serves notice on the other party. Either party should be allowed to take the steps required to rectify the problem and be allowed reasonable time to take such action.

Application to Scotland and Northern Ireland (clause 12)

Where the contract is to be constructed in Scotland, then Scottish law applies.

Construction (Design and Management) Regulations 1994

This clause has been added to the second edition to comply with the above regulations.

Appendix

The Appendix to the Conditions of Contract includes the following;

1. A short description of the wroks
2. Basis of payment. This may include any or all of the following on a single contract (clause 7):
   - lump sum
   - measure and value using a bill of quantities
   - schedule of rates
   - daywork schedule
   - cost plus
3. Where appropriate the method of measurement used
4. Name of engineer (clause 2.1)
5. Starting date (clause 4.1)
6. Period for completion (clause 4.2)
7. Period of completion for different sections of the work (clause 4.2)
8. Amount of liquidated damages (clause 4.6)
9. Limit of liquidated damages (clause 4.6)
10. Defects correction period (clause 5.1)
11. Rate of retention (clause 7.3)
12. Limit of retention (clause 7.3)
13. Minimum amount of interim certificates (clause 7.3)
14. Bank whose base lending rate is to be used (clause 7.8)
15. Insurance of the works (clause 10.1)
16. Minimum amount of third party insurance (persons and property) (clause 10.6)

Guidance notes

There are currently thirteen guidance notes. These are not part of the contract but have been provided to assist the users with the administration of the contract. The following are some of the points that are included:

- Appendix to the contract should be prepared prior to inviting tenders.
- No provision is made for price fluctuations, due to the short duration of contracts.
- Access to the site should be available at the starting date under clause 4.1.
- The procedures for letting and administering a minor works contract are intended to be as simple as possible.

Conciliation Procedure (1994)

These are the procedures to be used to resolve difficulties prior to arbitration or legal procedures being initiated.

CHAPTER 20

FCEC FORM OF SUBCONTRACT

Introduction

The former Federation of Civil Engineering Contractors (FCEC) has prepared a form of subcontract for use with the ICE Conditions of Contract, sixth edition on civil engineering projects. The current edition of this form was prepared in 1991. It comprises the formal agreement, twenty clauses and five schedules.

Agreement

This is the formal agreement for completion and signing by the two parties concerned, i.e. the main contractor and each individual subcontractor.

Clauses

Definition (clause 1)

This clause includes definitions that are appropriate to the subcontract, some of which are included in the main contract conditions. These include:

- Main contract: the particulars which are given in the First Schedules in the subcontract
- Subcontract: an individual subcontract. The documents are referred to in the Second Schedule. These exclude any standard printed conditions of a subcontractor
- Subcontract works: those that are described in the Second Schedule
- Main works: the works of the main contract
- Price: the sum specified in the Third Schedule, that is payable to a subcontractor for the subcontract works

General (clause 2)

These specify the subcontractor's obligations to:

- execute, complete and maintain the subcontract works in accordance with the subcontract and to the reasonable satisfaction of the contractor and engineer
- exercise reasonable skill, care and diligence in designing any part of the subcontract works
- provide all labour, materials, subcontractor's equipment and temporary works that are required for the subcontract works
- not assign or sublet the works without the written consent of the contractor
- note that copyright of the contract documents will not pass to the subcontractor.

Main contract (clause 3)

The subcontractor will be deemed to have full knowledge of the provisions of the main contract, other than details of the main contractor's pricing. The main contractor must provide a subcontractor with a copy of the appendix to the main contract if requested, together with details of any non-standard ICE Conditions of Contract requirements. There is no privity of contract between the subcontractor and the employer. The subcontractor must indemnify the contractor against every liability which the main contractor may possibly incur to other persons. These may include claims, demands, proceedings, damages, costs or expenses by reason of a breach by a subcontractor. A subcontractor must also acknowledge that any breach may result in the main contractor becoming liable under the main contract. Such damages may thus become the responsibility of a subcontractor.

Contractor's facilities (clause 4)

The main contractor must allow subcontractors the use of, for example, standing scaffolding, i.e. scaffolding normally provided by the contractor for his own use. However, the contractor does not need to retain this solely for the use of a subcontractor. The main contractor's permission does not imply any warranty for its fitness, condition or suitability. The main contractor is not liable to a subcontractor regarding its use. The main contractor must allow subcontractors the use of common facilities as outlined in part 1 of the Fourth Schedule. Facilities exclusive to individual subcontractors must also be provided in accordance with part 2 of the Fourth Schedule. These exclusive facilities are amenities that are required by some subcontractors. They may include the use of some plant or machinery owned by the main contractor or assistance by the main contractor with unloading or transportation around the site of a subcontractor's goods or materials. The subcontractor's responsibility is to indemnify the main contractor against damage or loss from misuse by the subcontractor of the main contractor's equipment or other facilities.

Site working and access (clause 5)

Subcontractors must generally observe the same site hours as the main contractor. They must comply with the reasonable rules and regulations adopted by the main contractor in the execution of their work, the delivery and storage of materials and equipment. The main contractor must make available parts of the site, including access, to allow subcontractors to carry out their work. The contractor is not bound to give exclusive access or possession to an individual subcontractor. Subcontractors must permit reasonable access to the engineer or the engineer's representative to the subcontract works. The main contractor and other subcontractors must also be allowed reasonable access in order to allow them to perform their own duties and responsibilities on the site.

Commencement and completion (clause 6)

A subcontractor should commence work on site within 10 days, or other period that is agreed in writing, of receipt of the contractor's written instructions. A subcontractor should thereafter proceed diligently to complete the subcontract works. The subcontractor should complete the works as specified in accordance with the period for completion specified in the Third Schedule. A subcontractor may be delayed in the execution of a subcontract for one of the following reasons:

- circumstances that entitle the main contractor to an extension of time (where this is appropriate or affects the subcontract works)
- the ordering of a variation to the subcontract works
- a breach by the main contractor of the subcontract

Under these circumstances the subcontractor may be entitled to an extension of the period of completion that is fair and reasonable. If different periods of completion are stated in a subcontract (Third Schedule), then for the purpose of granting an extension of time each part will be treated separately. The subcontractor is able to commence work off-site, for instance for fabrication purposes, at any time prior to receiving the contractor's written instructions described above. The contractor must notify a subcontractor, in writing, where an extension of time has been granted.

Instructions and decisions (clause 7)

Subcontractors must comply with the instructions from the engineer and the engineer's representative. These instructions are notified and confirmed in writing by the main contractor. A subcontractor has the same rights to payment from the main contractor as the main contractor has from the employer. Where instructions or decisions are given incorrectly by the engineer under the main contract, then a subcontractor is entitled to recover reasonable costs from the contractor in complying with such instructions. The main contractor has similar powers in relation to a subcontract, regarding instructions and decisions, as the engineer has under the main contract. These powers can be used, even where the engineer has not used them under a main contract.

Variations (clause 8)

Subcontractors must make variations to the subcontract works, through addition, modification or omission where:

- they are ordered by the engineer under the main contract and confirmed in writing to the subcontractor from the main contractor
- they are agreed by the employer and contractor and confirmed in writing to the subcontractor from the main contractor
- they are ordered in writing by the main contractor

Any order, relating to a subcontract, given by the engineer under the main contract constitutes a variation where it is subsequently confirmed in writing by the main contractor. Oral instructions from the engineer or employer are not to be acted upon by a subcontractor unless they are confirmed in writing from the main contractor. Subcontractors must not make any alteration or modification to the subcontract works. Variations to subcontracts are valued under clause 9 (Valuation of variations) and paid for under clause 15 (Payment).

Valuation of variations (clause 9)

All authorised variations to subcontract works are valued in accordance with the rules for the valuation of variations that are used in the main contract (see Chapter 17). The price specified in the Third Schedule of the subcontract is adjusted for any addition or deduction. Where the subcontract is based upon a lump sum price, the changes to the subcontract works are to be valued on a fair and reasonable basis. Subcontractors are invited to attend at any remeasurement of their work that may take place between the engineer and the main contractor. Where the work is to be valued on a daywork basis, either the schedule included in the bill of quantities is to be used or the Schedule of Dayworks carried out incidental to Contract Work (issued by the former Federation of Civil Engineering Contractors).

Notices and claims (clause 10)

The main contractor is sometimes required to provide a return, account, notice or other information to the engineer or employer. Where this information relates to subcontract works, then the main contractor may require this to be provided by a subcontractor to enable the return to be made to the engineer or employer. Subcontractors must be properly informed of this requirement. The main contractor must take all reasonable steps to secure from the employer any contractual benefits that may affect the execution of any subcontract works. If the main contractor receives contractual benefits from the employer, such as an extension of time, this should, in turn, be passed on to subcontractors, where this is fair and reasonable. Where a subcontractor fails to comply with the requirements of this clause and this prevents a main contractor from recovering monies from the employer, the main contractor may deduct such amounts from the subcontractor concerned.

Property in materials and plant (clause 11)

The property of the subcontractor, i.e. equipment, temporary works, materials, etc. is vested in the employer or revested in the main contractor during the period of the contract.

Indemnities (clause 12)

A subcontractor must indemnify the main contractor against all liabilities to other persons for bodily injury, damage to property or other loss arising from the execution of subcontract works. This indemnity must cover all costs, charges and expenses that may be involved. The contractor cannot claim for such losses from both the employer and a subcontractor. The subcontractor is not liable for any losses caused by the wrongful acts of the main contractor or other persons who are employed by the contractor. The main contractor must indemnify the subcontractor against all liabilities and claims against which the employer undertakes to indemnify the main contractor.

Outstanding work and defects (clause 13)

Subcontractors must complete their work prior to the substantial completion of the main works. Where the main contract works are completed in sections, the subcontract works should be completed before the substantial completion of these sections. Subcontractors must maintain their work to the satisfaction of the engineer and make good all defects and imperfections. Where these defects or imperfections are due to the negligence of a subcontractor, they must be made good at the subcontractor's expense. Subcontractors are liable for their work, under the contract, until the end of the defects correction period. Where making good is required as a result of negligence on the part of the main contractor or those for whom the main contractor is responsible, then a subcontractor is entitled to receive payment from the contractor for the costs of making good. Where the defect is due to damage caused by another subcontractor, then the main contractor would offset the costs involved against such a subcontractor.

Insurances (clause 14)

Subcontractors must comply with the insurance requirements outlined in Part 1 of the Fifth Schedule which are the same as those contained in the main contract. In order to avoid insuring the works twice, the responsibility for these is usually undertaken by the main contractor. Where the subcontract works suffer an insurable damage, a claim will be made against the main contractor's insurance. Where damage occurs to a subcontractor's plant or equipment, the subcontractor would normally claim from his own insurance company. A requirement of the main contract is that the main contractor must maintain insurance of the works until the issue of the certificate of substantial completion and thereafter for those aspects covering defects correction. This insurance will be maintained even though the main contractor's work has been completed and only subcontractors' work is being carried out.

Payment (clause 15)

The following procedures are adopted for payment to subcontractors:

1. The main contractor will have agreed the monthly dates (specified date) for interim payments.
2. Subcontractors must submit their own written statements to the contractor within seven days of those dates. These statements must indicate:
   * the value of all work properly executed under the subcontract
   * materials delivered to site
   * materials off-site, where these are allowable under the terms of the subcontract agreement
3. The statement must be in a suitable form and contain such information as required by the contractor.
4. The value of the work done is calculated on:
   * the quantities of work completed using the bill rates
   * a proportion of the lump sum price
5. A subcontractor's statement constitutes a valid statement for this purpose only.
6. The main contractor will make an application for payment using the valid statement.
7. Where a contractor seeks to enforce payment from the employer, unpaid monies owed to subcontractors must be included in such applications.
8. A subcontractor should receive payment within 35 days of the specified date. The amount due is that agreed in the valid statement less previous payments and the retention.
9. Subject to:
   * clause 3 (Breach of subcontract)
   * clause 10 (Contractual claims)
   * clause 17 (Subcontractor's default)

the main contractor is entitled to withhold or defer payments due to a subcontractor.

10. Payments will also be withheld where:
    - the amount of a subcontractor's payment does not justify the issue of an interim certificate
    - the amount, and that of other subcontracts and the main contractor, is insufficient to justify the issue of a certificate by the engineer under the main contract
    - the amounts included in the statement are not certified in full by the engineer
    - the employer has failed to pay an engineer's certificate in full
    - a dispute has arisen between the subcontractor and contractor, or the contractor and employer

11. Where a contractor withholds payment from a subcontractor, the subcontractor must be informed of the reasons in writing.

12. In the event of contractor failing to make a payment at the appropriate time, a subcontractor is entitled to interest at the rate agreed in the main contract.

13. Notwithstanding 12 above, a subcontractor must be paid interest actually received by the contractor from the employer which is attributable to monies due to a subcontractor.

14. When the main contractor receives the release of the retention monies (first and second release), subcontractors should receive their proportion of these amounts.

15. Final payment to a subcontractor should be made within 3 months of a subcontractor completing his obligations, including outstanding work and defects or within 14 days of the main contractor receiving the final payment under the main contract.

16. The main contractor is not liable to a subcontractor after the issue of the defects correction certificate in respect of the main contract works, unless the subcontractor has previously made a claim in writing to the main contractor.

Determination of the main contract (clause 16)

If the main contract is determined this usually results in the determination of all subcontracts. Under clause 11 (Property in materials) a subcontractor will be required to remove all men and equipment from the site. Depending upon the reason for the determination of the main contract, a subcontractor may be re-engaged to complete the works under a separate agreement. Upon such a determination a subcontractor is entitled to be paid:

- value of work properly completed
- materials on site
- reasonable costs of removing subcontractor's equipment from site

- fabricated goods and materials off-site (to be subsequently delivered to the site)

If the main contract is determined as a consequence of a breach by a subcontractor, the provisions above will not apply.

Subcontractor's default (clause 17)

The determination of a subcontractor may occur because the subcontractor:

- fails to proceed with the works with due diligence
- fails to perform the works in accordance with the subcontract
- refuses or neglects to remove defective materials
- refuses or neglects to make good defective work
- becomes bankrupt or goes into liquidation
- is required to be removed from site by the engineer

Under these circumstances the main contractor may use a subcontractor's materials, equipment, etc., for the purpose of completing the subcontract works. Upon such determination, the rights and liabilities of the contractor and a subcontractor are the same as the preceding clause. Alternatively, the contractor may, in lieu of giving notice of determination to a subcontractor, complete the subcontract works and deduct any costs involved from monies outstanding to a subcontractor.

Disputes (clause 18)

Any dispute between the main contractor and a subcontractor is to be settled in accordance with the following provisions:

- The parties involved must be informed of a dispute and allowed a reasonable time to take appropriate action.
- Within 28 days of the service of a notice of dispute, the parties must give notice under the ICE Conciliation Procedure (1988).
- Within 28 days of the receipt of the conciliator's recommendation, the parties must refer the dispute to arbitration.
- The notice of dispute may within 28 days be referred directly to arbitration.
- If the parties cannot agree upon the appointment of an arbitrator, the dispute will be referred to the president of the ICE.
- Any reference to arbitration will be in accordance with the ICE Civil Engineer's Arbitration Procedure (1983).

Where a dispute involves the main contract, the subcontractor may be requested to attend meetings and to provide information as requested by the main contractor.

Value Added Tax (clause 19)

Subcontractors' rates and prices are deemed to be exclusive of any VAT.

Law of subcontract (clause 20)

The law of the country applying to the main contract will apply to a subcontract.

Schedules

There are five schedules to be completed. Accompanying these schedules are notes for the guidance of contractors on the completion of the schedules. The first three schedules describe the particulars from the main contract.

First schedule

The first schedule should identify the main contract and specify the documents of which it consists. These include:

- the Agreement and the date when this was signed between the employer and the contractor.
- general conditions of contract (normally the ICE sixth edition)
- specification
- bills of quantities

This schedule should include a brief description of the works that are to be constructed. The schedule will also indicate the dates when the contractor will submit statements to the engineer for interim payments under the main contract. Where these dates are not known, subcontractors must be informed of them as soon as possible. This is to allow them sufficient time to submit their invoices to the contractor for inclusion in the statement of work done. The minimum amount of interim certificates under the main contract should also be stated (clause 60 (3)).

Second schedule

This schedule identifies three aspects of the subcontract.

1. This is concerned with any further documents that will form a part of the subcontract. These may include, for a particular subcontract:
   - priced bills of quantities
   - drawings
   - specifications
   - schedules of rates or prices
   - quotations
   - estimates

Where the subcontract works have been fully described in the main contract documents, extracts from these documents, e.g. bills of quantities, specifications, etc. should be listed. The extracts would not normally include the rates and prices inserted by the main contractor, since these are between the employer and the main contractor. Separate rates would normally be provided between the main contractor and a particular subcontractor. Where information from a subcontractor is included, such as quotations, care should be exercised to ensure that none of the contract terms that might be listed is in conflict with the main contract conditions. Conditions that are inconsistent with the main contract should be deleted. In some circumstances it may be necessary to incorporate special conditions. These could be particularly onerous on the main contractor and should be avoided unless they can be transferred to the employer. Where the subcontract works are to be carried out for a lump sum, not based upon measured quantities, the subcontract should clearly state the extent to which the work is provisional and whether its value will or will not be recalculated.

2. The full extent of the subcontract works should be described. These should clearly define the materials to be supplied and the work to be executed.
3. If the subcontract is to be subject to any contract price fluctuations, sufficient detail should be given to make them clear.

Third schedule

This schedule is also in three parts.

1. The price is to be deleted if the contract is on a measure and value basis. Alternatively, if the price is calculated on a lump sum basis, measure and value should be deleted.
2. The percentage retained by the employer from interim certificates must be stated and may be different for the works completed and the materials on site. If there is a limit to the amount of retention, whether as a sum of money or a percentage, this must be stated.
3. The period for completion should be stated in weeks or months. Where it is appropriate, the start and completion dates for the particular subcontract should be provided.

Fourth schedule

The fourth schedule is concerned with the use of the main contractor's facilities as described in clause 4. If the contractor proposes to charge a subcontractor, the schedule should specify the charges involved. The items that are to be provided are separately described as:

Common facilities (clause 4 (2)): These include water, lighting, use of main contractor's equipment, which the contractor is prepared to allow a subcontractor to use. Common facilities normally include those already provided by the main contractor, such as the use of standing scaffolding. They may also include the use of the main contractor's cranes for lifting purposes.

Exclusive facilities (clause 4 (3)): These are facilities provided to an individual subcontractor.

Fifth schedule

This schedule covers matters relating to insurances and to clause 14 of the subcontract conditions. There are two parts of the schedule to be completed. The intention is to ensure that the requirements of the main contract are complied with and that there is no unnecessary duplication of insurance.

Part 1 specifies insurances effected by a subcontractor

Part 2 specifies the policy of insurance which the contractor has effected in pursuance of clause 21 of the main contract conditions. Where a subcontractor is not entitled to benefit under the policy of the main contractor, that part should be marked as not applicable.

CHAPTER 21

ICE DESIGN AND CONSTRUCT CONDITIONS OF CONTRACT

Introduction

The ICE Design and Construct Conditions of Contract were introduced in 1992 to reflect the changing needs of the construction industry and the new demands being introduced by the industry's employers. The Conditions are sponsored by the Institution of Civil Engineers, the Association of Consulting Engineers and the Federation of Civil Engineering Contractors.

The emphasis of these conditions is towards single-point responsibility as far as the employer is concerned. This removes any disagreement over difference of opinion between the engineer and the contractor. In some cases the contractor will blame poor performance on an incomplete design. In others, the engineer may attribute a problem to the contractor failing to carry out the works in a workmanlike manner. The use of design and construct generally removes such disagreements, by allowing the design and construct contractor to solve the problems internally without involving the employer. It is generally assumed, however, that it is less easy for an employer to change a contractor's design and that doing so often involves considerable expenditure.

Much of the layout and content of the Conditions is similar to the ICE Conditions of Contract (sixth edition) and the clauses and their conditions reflect this similarity. Notwithstanding these similarities and origins, this new contract is not a version of the ICE main form of contract. The following are some of its salient points.

Definitions and interpretation (clause 1)

An employer's representative is introduced to act as such for the purposes of the contract. In practice the employer's representative is likely to be a chartered engineer, although this need not necessarily be the case. The contractor's representative adopts the dual role of designer and constructor and consequently the significance and importance of this appointment is greater than that of the contractor's representative under the ICE main form.

The employer's requirements are a combination of design and construct requirements.

Employer's representative (clause 2)

The employer's representative replaces the engineer under the main form but, due to the nature of the work, has far fewer responsibilities and activities to perform. The representative can delegate responsibilities to others. When issuing instructions to the contractor on behalf of the employer, these must be in writing.

Whilst the engineer acts with a certain amount of impartiality under the main ICE form (clause 2 (8)) this is not the case under the Design and Construct Conditions.

Assignment and subcontracting (clauses 3–4)

These clauses follow closely the provisions of the main ICE Conditions of Contract.

Documentation and information (clauses 5–7)

The contract documents are not defined. With the exception of the conditions of contract the documents may vary depending upon the type and nature of the project to be constructed. The actual documents that form the contract are to be listed on the form of tender. They will include, as a minimum:

- employer's requirements
- conditions of contract
- designs and drawings
- specifications of quality and standards
- methods of calculating costs and payments

The documents are to be mutually exclusive. Where ambiguities or discrepancies are discovered in the employer's requirements, these will be adjusted accordingly by the employer's representative. Where the contractor requires further information about the design or construction of the works, this will usually be sought from the employer's representative.

The employer is responsible for supplying two complete sets of the contract documents to the contractor after the award of the contract. Copyright in the project remains with the party who prepared the information. This means, for example, that copyright in the design and specification, prepared by the contractor, remains with the contractor and does not pass to the employer.

General obligations (clauses 8–35)

The contractor is responsible for the design and construction of the works. The contractor must exercise all reasonable skill, care and diligence in the execution of these duties. The contractor must submit a quality plan and procedures for the prior approval of the employer's representative. However, compliance with such a plan will not necessarily relieve the contractor from other duties, obligations or liabilities under the contract. Where Acts of Parliament, regulations or bye-laws require checks or tests on the design or construction work, these are the contractor's responsibility.

The contractor is deemed to have visited the site to determine:

- the nature of the ground and subsoil
- the extent and nature of work and materials that are necessary for construction
- communication and access to the site

The contractor is deemed to have based the tender on the information made available by the employer and on the inspection and examination of the site. The contractor can only do what is reasonable under the circumstances and, like many civil engineering contracts, design and construct projects will encounter problems during their construction. Where problems occur that could not reasonably have been foreseen by an experienced contractor, the contractor may be entitled to additional payments.

The contractor should, within 21 days of the award of the contract, submit to the employer's representative a copy of the contractor's programme. The employer's representative should, within a further 21 days, either approve the programme or request that it be modified. The contractor should then seek to ensure that progress is made in accordance with this programme.

The contractor must submit details of the design and construction methods for the approval of the employer's representative.

Materials and workmanship (clause 36)

The contractor must submit to the employer's representative for approval, proposals for checking the design and setting out of the works and testing of materials and workmanship. This is in accordance with the contractor's quality assurance system outlined in clause 8 (3). The contractor must carry out tests approved in the contract and other tests that may be reasonably required by the employer's representative.

Commencement and delays (clauses 37–46)

These clauses follow closely the provisions of the main ICE Conditions of Contract.

Liquidated damages for delay (clause 47)

These clauses follow closely the provisions of the main ICE Conditions of Contract.

Certificate of substantial completion (clause 48)

Prior to the issue of the certificate of substantial completion of the works, the contractor must provide the employer with operation and maintenance instructions. These must be in sufficient detail to enable the works (or section) to be operated, maintained, dismantled, reassembled and adjusted satisfactorily.

Outstanding work and defects (clauses 49–50)

These clauses follow closely the provisions of the main ICE Conditions of Contract.

Alterations and additional payments (clauses 51–53)

The employer can request variations. These must be done in consultation with the contractor. Upon the receipt of a variation instruction from the employer's representative the contractor should submit:

- a quotation for the work necessitated by the variation
- an estimate of any delay
- an estimate of the costs of any delay

The employer's representative may accept these submissions or negotiate what is considered to be a fair and reasonable agreement. The contract price is then amended accordingly. The contractor may consider that changes to the works or conditions under which the work is executed require additional payments. Notice of an intention to claim should be submitted to the employer's representative within 28 days of an event occurring. Any amounts that can be agreed are included in interim payments to the contractor.

Materials and contractor's equipment (clause 54)

These clauses follow closely the provisions of the main ICE Conditions of Contract.

Measurement (clauses 55–57)

Since the contractor has provided the design and specification, all errors occurring in the documentation or pricing are the responsibility of the contractor. The contractor can choose to calculate the costs of the project in any way. There is no requirement to prepare a bill of quantities or to adopt a formal method of measurement. Of course, if the employer requires this document to be provided, this can be specified.

Prime cost items (clauses 58–59)

Since these are not nominated by the employer's representative, they are adjusted in accordance with clause 52. However, their financial adjustment will follow closely the adjustments that are made normally to prime cost sums. A percentage to be applied for overheads and profit in adjusting prime cost sums is to be included in the appendix.

Certificates and payments (clauses 60–61)

These clauses follow closely the provisions of the main ICE Conditions of Contract.

Remedies and powers (clauses 62–65)

These clauses follow the provisions of the main ICE Conditions of Contract. They also include the clauses covering frustration and war, which are provided for separately in the ICE Conditions.

Settlement of disputes (clause 66)

These clauses follow closely the provisions of the main ICE Conditions of Contract.

Application to Scotland etc. (clause 67)

These clauses follow closely the provisions of the main ICE Conditions of Contract.

Notices (clause 68)

These clauses follow closely the provisions of the main ICE Conditions of Contract.

Tax matters (clauses 69–70)

These clauses follow closely the provisions of the main ICE Conditions of Contract.

Special conditions (clause 71)

These clauses follow closely the provisions of the main ICE Conditions of Contract.

APPENDICES

TABLE OF RELEVANT STATUTES

A complete list of statutes (Acts of Parliament) can be found in Halsbury's *Statutes of England* (4th edn).

Administration of Justice (Appeals) 1934
Ancient Monuments Act 1931
Arbitration Acts, 1934, 1950, 1975, 1979
Architects (Registration) Acts, 1931, 1938
Architects Registration (Amendment) Act 1969

Bankruptcy Acts, 1869, 1914
Building Act 1984
Building Societies Act 1986

Civil Jurisdiction and Judgments Act 1982
Companies Acts, 1948, 1985, 1989
Consumer Credit Act 1974
Consumer Protection Act 1987
Contracts (Applicable Law) Act 1990
Control of Pollution Act 1974
Copyright, Design and Patents Act 1988
County Court Act 1984
Courts and Legal Services Act 1990

Defective Premises Act 1972

Employer's Liability (Compulsory Insurance) Act 1969
Employment Protection Acts, 1975, 1978
Environmental Protection Act 1990
European Communities Act 1972

Factories Acts, 1937, 1961
Fair Trading Act 1973
Finance Acts, 1970, 1971, 1972, 1975, 1980, 1985, 1989
Financial Services Act 1986

Health and Safety at Work Act 1974
Highways Acts, 1959, 1980
Housing Acts, 1936, 1957, 1959, 1961, 1974, 1985

Income and Corporation Taxes Act 1988
Industrial Training Act 1964
Insolvency Act 1986

Landlord and Tenant Act 1927
Latent Damage Act 1986
Law of Property Acts, 1925, 1989
Limitation Acts, 1939, 1963, 1980
Local Government Acts, 1933, 1972, 1988
London Building Acts, 1894, 1930, 1939

Magistrates' Court Act 1980
Mines and Quarries Act 1954
Misrepresentation Act 1967

National Heritage Act 1983
National Insurance Act 1965

Official Secrets Acts, 1911, 1939

Partnership Act 1890
Patents Acts, 1949, 1977
Planning and Compensation Act 1991
Planning (Listed Buildings and Conservation) Act 1990
Prevention of Corruption Acts, 1889, 1916
Public Health Acts, 1848, 1875, 1908, 1936, 1961
Public Utilities Street Works Act 1950

Race Relations Act 1976
Restrictive Practices Act 1976
Rights of Light Act 1959
Road Traffic Act 1930

Sale of Goods Acts, 1893, 1979

Town and Country Planning Acts, 1947, 1968, 1971, 1990
Trade Descriptions Act 1968

Unfair Contract Terms Act 1977

War Damage Acts, 1943, 1965

CHRONOLOGY OF A CONTRACT UNDER THE ICE CONDITIONS OF CONTRACT

Activity (Certificates in bold)	Timing	Clause
Works commencement date		41 (1)
Time for completion	Contract period	43
Programme supplied by contractor	Within 21 days of the award of the contract	14
Possession of the site by contractor		42
Start on site	Within a reasonable period	41 (2)
Rate of progress		46
Interim certificates	See Appendix 3	60
Extension of time		44
Suspension of works		40
Certificate of substantial completion of parts of the works		48 (4)
Certificate of substantial completion		48
Defects correction period		49
Defects correction certificate		61
Clearance of site on completion		33
Final account	3 months after the issue of the defects correction certificate	60 (4)
Final certificate		60 (4)

APPENDIX 3

COST CONTROL

The importance of control of construction costs has increased considerably in recent years. This has been due to the need to be more accountable in both the public and private sectors and the desire to achieve value for money. There are also few employers who have access to a limitless supply of money and it is important that they should be fully aware of the cost implications of their capital works projects before they commit themselves to the expenditure.

The financial control of any construction project commences at inception and continues until the issue of the final certificate. The method used for controlling costs will depend upon the following circumstances:

- the method used for contractor selection
- the method used for price determination, both for the tender sum and the final account
- whether the control is being exercised for the contractor or the employer
- the role and relationship of the quantity surveyor or measurement engineer in respect of budgeting and accounting

Whilst the employer's and contractor's needs are similar, their objectives are different. Both wish to restrain their costs. However, the contractor's aim is to increase their profits, whilst that of the employer is to secure the highest possible value for money within budget.

Pre-contract

The process of cost control starts at the inception of the scheme, usually with some form of outline budget. This is necessarily imprecise since neither site investigations nor the design of the works will have been considered in any great detail. Throughout the design phase, as the project begins to take shape, the appropriate costs of the design should be properly monitored through a process of cost planning. This process initially analyses the budget under the different elements or aspects of the project. On a new highway, for example, subdivisions of the overall cost may be allocated to roads, bridgeworks, drainage, accommodation works, etc. It may also be desirable to further allocate the bridgeworks costs to individual bridges or even the broad components that go together to make a bridge structure (abutments, piers, deck, etc.). As the design develops a close control can be kept on the cost implications of the design.

Using such a system should ensure that the engineer and the employer are kept fully informed of the cost implications of the design, where this can be known. It should result, if carefully and competently carried out, in the forecast of contractor's tenders showing no surprises and being acceptable to the employer.

Where tender sums are much higher than otherwise expected or indicated, certain aspects of the design may need to be reconsidered. This results in abortive work and the possibility of the design becoming unbalanced in terms of its aesthetics or quality.

Post-contract

Once the Agreement has been signed between the employer and the contractor, the project moves into its post-contract phase. The employer will have an agreed tender sum that is now much more precise than the pre-contract budget and estimate checks. Civil engineering projects are more difficult to control financially than building projects, since much of the work is at or below ground level. Until the contractor actually begins work on site, not even the best site investigation can state with certainty what difficulties might be encountered.

The engineer and the employer must be kept fully informed of any changes to the costs involved. The interim certificates will measure what has been expended but will not provide an overall indication of the final costs of the project.

The essence of cost control is the frequent and regular reporting to the employer of a financial statement of the contract. It will be necessary to assess the cost consequences of engineer's instructions in changes to the design, or of changes necessitated by the site conditions. The employer will need to be aware of any likely changes to the prospective financial commitment prior to the execution of the work on site. This demands a close working relationship between the engineer and the quantity surveyor or measurement engineer.

Post-contract cost control includes the following:

- interim valuations and certificates for payments
- cashflow control and forecasts through budgetary control
- financial statements showing the current and expected final costs for the project
- final account, the agreement of the final certificate and the settlement of claims

Throughout the post-contract period the employer will be informed of the expected final cost of the project through the issue of financial statements.

The responsibilities in respect of post-contract cost control will vary, depending upon the terms of appointment and the provisions within the contract conditions. The following are some of the activities involved:

- attendance at site meetings
- preparation of documentation for subcontractors and suppliers, examination of quotations and invoices and making recommendation
- advising on contractual implications
- negotiation; measurement and unit rates
- confirmation of payments to nominated subcontractors
- advising on the financial implications of extensions to the contract period
- preparation of special reports or cost implications
- completion of documentation which may be required for some clients, particularly government departments
- working with auditors

Employer's financial reports

Financial statements are prepared at regular times throughout the contract period and in sufficient detail to provide an indication of the costs already committed and the final cost of the project. The size and complexity of the project will determine how often they are prepared, e.g. four times per year. The report is in two parts: the first part considers the current position and the second part the likely final cost of the project.

The current position will state how much has already been expended in terms of interim payments, including retention. This sum will be compared with an expenditure forecast plan to assist the employer's cash flow.

The second part of the report can be prepared in different ways and levels of detail, depending on who will use it. Most employers will want to know their overall anticipated financial commitment. The report therefore seeks to forewarn the employer of any future possible increases or decreases in the costs of the project. The technical details, even on a simplified report, will mean very little to the typical employer. However, the engineer may require more information in order to plan and rectify the remaining expenditure for the project.

The financial statement updates the tender sum. The accuracy of the amounts included will vary, depending upon the quality of the judgement provided. The financial statement will include:

- Quotations set against prime cost and provisional sums already accepted although the work may still be incomplete or not even started.
- Variation orders which have been issued to the contractor. The engineer should seek cost advice prior to the issue of any instruction.
- Re-measurement of the work, even though this may represent only an approximate calculation and may still have to be agreed with the contractor.

- Daywork sheets are often not submitted as requested, but some allowance needs to be made for them by way of an estimate.
- Claims for delays or disruptions may already have been intimated, and it would be foolish to discount these from the statement. An amount will need to be included, together with an assessment of the likelihood of their acceptance.
- Where the contract includes the contract price fluctuation clause, the effects of increased costs will need to be taken into account.
- Where other changes are being anticipated, these should also be taken into account in the report.

Further information can be found in A. Ashworth, *Cost Studies of Buildings* 2nd edn., 1994, Longman.

APPENDIX 4

ENGINEER'S INSTRUCTIONS

The engineer has wide powers to issue instructions under the ICE Conditions of Contract. These Conditions also limit the type of instructions that can be given to the contractor. Where the contractor thinks that the engineer is exceeding the powers provided under the Conditions, then the contractor can require the engineer to justify the authority for issuing a particular instruction.

The contractor normally receives instructions from the engineer named in the contract, or, subject to certain limitations, from the engineer's representative. In order to be properly valid, these instructions from the engineer to the contractor must be in writing. Instructions given orally must be later confirmed in writing.

The following clauses refer to instructions that may be given by the engineer to the contractor:

Clause 2 (6)	Instructions in writing
Clause 2 (6)	Confirmation of oral instructions
Clause 2 (6)	Power to issue instructions
Clause 4 (5)	Removal from the works of a subcontractor
Clause 5	Discrepancies in contract documents
Clause 7	Issue of modified drawings or specifications
Clause 7	Failure to issue drawings, specifications or instructions
Clause 7	Approval of permanent works designed by the contractor
Clause 12	Receipt of a contractor's claim
Clause 14 (2)	Approve the contractor's programme
Clause 14 (4)	Require the contractor to provide a revised programme
Clause 14 (5)	Provide design criteria for permanent or temporary works
Clause 14 (6)	Require information pertaining to methods of construction
Clause 15 (2)	Instructions given to the contractor's agent
Clause 16	Removal of any person employed on the works
Clause 17	Action to be followed on incorrect setting out by the contractor
Clause 18	To make boreholes or exploratory excavations
Clause 19	Aspects of safety and security
Clause 20 (3)	Apportion costs arising from damage due to an excepted risk
Clause 26 (3)	Instructions must conform to Acts, regulations or bye-laws

Clause 31 (1) Facilities for other contractors
Clause 32   Discovery of fossils
Clause 35   Returns of labour and contractor's equipment
Clause 36   Quality of materials and workmanship
Clause 36   Tests on workmanship and materials
Clause 38 (1) Examination of work before covering up
Clause 38 (2) Uncovering and making openings
Clause 39 (1) Removal of unsatisfactory work and materials
Clause 39 (2) Default of the contractor in compliance with instructions
Clause 40   Suspension of the works
Clause 41   Works commencement date
Clause 42 (3) Failure of the employer to give possession of the site
Clause 44   Determine an extension of time
Clause 45   Night and Sunday work
Clause 46   Rate of contractor's progress
Clause 47   Assessment of liquidated damages
Clause 48   Outstanding work to be completed
Clause 48   Outstanding work and defects to be made good
Clause 50   Contractor to carry out searches
Clause 51   Issue variations
Clause 52   Value variations
Clause 52 (4) Response to contractor's claims
Clause 53   Consent for contractor to remove from site contractor's equipment, etc.
Clause 55   Correction of errors in contract documents
Clause 58 (1) Ordering of provisional sums
Clause 58 (2) Ordering of prime cost sums
Clause 59 (2   Engineer's action upon contractor's objection to nominated subcontractor
Clause 59 (4) Engineer's action on the termination of a nominated subcontract
Clause 62   Urgent works
Clause 63   Determination of the contractor's employment
Clause 64   Frustration of contract
Clause 65   Outbreak of war or hostilities
Clause 66 (3) Settlement of disputes

APPENDIX 5

INTERIM CERTIFICATES

Considerable amounts of capital are required by contractors to finance projects of civil engineering construction. The costs of raising this capital can be expensive to the contractor and payment is therefore usually made on account. Payments for works of civil engineering may be made:

- at certain stages of the contract
- when the work has been fully completed
- in advance of any of the work commencing
- as the work proceeds

The last of these methods is the one that is the most commonly employed on civil engineering projects.

Purpose

To assist the contractor's cash flow and thus reduce the borrowing requirements that would otherwise need to be added to a contractor's tender.

Some consideration should be given where the work is not on schedule or the rates have been unbalanced by the contractor.

The issue of a certificate does not confirm the acceptance of the quality or standards of work.

Employer

Within 28 days of receiving the contractor's statement, the engineer must certify the agreed value and the employer make the appropriate payment to the contractor.

Contractor

The payment made is based upon the contractor's assessment of the work completed. This is approved, where appropriate, by the engineer. From the amount received under interim certificates, the contractor will make payments to the contractor's subcontractors and nominated subcontractors. In the case of default in making payments to nominated subcontractors, the engineer has the authority in the contract to make direct payments to such firms.

Contractor's entitlement

(See Figure A5.1 on page 290.)

- value of preliminary items
- value of permanent work completed
- relevant value of temporary work
- day works
- materials on site
- materials off-site at the engineer's discretion
- nominated subcontractor's work and materials
- increased costs, where the fluctuation clause applies
- value of any contractual claim

Retention

Interim payments are not paid in full, but a percentage of the amount certified is retained by the employer. This is deducted in case of default by the contractor and also to act as some incentive for the contractor to proceed with the works as quickly as possible and to complete the project.

It is recommended that the amount of retention should not exceed 5 per cent of the amount of the certificate with a limit of 3 per cent of the tender total.

The retention deductions are eventually paid to the contractor. One half is paid with the certificate of substantial completion of the works and the remainder with the issue of the defects correction certificate.

On some types of contract, the retention is paid into a joint, contractor and employer, account and paid to the contractor in the manner described above. Such a joint account may or may not generate interest.

Previous payments

Interim payments are valued gross for all the amounts that the contractor is entitled to at that time. Previous payments to the contractor are deducted from these amounts.

Frequency

Interim payments are usually prepared on a monthly basis but any period can be written into the appendix to the contract. On very large projects, a provisional amount may be made weekly to the contractor. Unless the monthly interim certificates reach the minimum amount stated in the contract, no payment to the contractor will be made during that period.

Value Added Tax

All interim certificates are exclusive of VAT.

Project...

Employer...

Engineer...

Interim Payment No .... Date.......................

Cumulative amount

Preliminaries

Measured works

Nominated subcontractors

Labour and other charges

Dayworks

Fluctuations in price

Contractual claims   _____

Materials on site   _____

Less Retention 3%   _____

Less Previous payments   _____

Amount due for payment   = = = = = =

Note: Some statements will include the previous valuation of the works to indicate the progress that has been achieved in that month.

Figure A5.1 Contractor's interim payment statement

APPENDIX 6

PRIME COST SUMS

Prime cost sums are amounts of money included in bills of quantities for construction work that will not normally be executed by the main contractor. They are frequently included to represent work that is of a specialist nature. This specialist work may involve the supply of materials, goods or components as well as work on site. The ICE Conditions of Contract make no distinction between nominated suppliers and nominated subcontractors, each are described as nominated subcontractors. The Civil Engineering Standard Method of Measurement (CESMM3) deals slightly differently with nominated subcontracts which include site work and those which are for the supply of materials, goods or components.

Clause 59 (5) of the ICE Conditions of Contract identifies the provisions for payment.

1. The amount for the prime cost sum is inserted in the bills of quantities during its preparation. It is normally based upon a quotation from a subcontractor. The engineer will usually obtain three competitive quotations. After the decision is made as to which quotation to accept a rounded up amount is included in the bills of quantities.
2. Before accepting a subcontractor's quotation, the engineer should ensure that the sum excludes trade and other discounts, rebates and allowances. A discount for prompt payment should be included, which the main contractor will retain before paying an invoice once the work has been executed. The ICE Conditions do not specify the amount of this discount, but it is usually assumed to be 5 per cent.
3. An amount can be inserted for labours by the main contractor when pricing the bills of quantities for tendering purposes. These really comprise two separate components, sometimes referred to as general and special attendance. The former is provided for all subcontractors and allows them to use the facilities already established on site by the main contractor. This item allows the main contractor the opportunity to add a charge for the use of such general facilities. Special labours or special attendance facilities are pertinent only to individual prime cost items. They usually require the main contractor to provide some sort of special assistance, such as hoisting or lifting. They should be described clearly to

enable the contractor to price them as accurately as possible. Where they are vague, the contractor may over-price the items. These labour items are priced as a lump sum and are only adjusted in a final account where the amount of work, rather than the price, involved has changed.

4. The main contractor is also given the opportunity of pricing other charges and profit. A typical percentage inserted by contractors is 5 per cent. The contractor will normally insert a percentage and the amount paid will be adjusted in the final account on the basis of a nominated subcontractor's invoice. The percentage is also stated in the Appendix to the form of tender.

5. Where the main contractor wishes to submit a quotation against a prime cost sum, the quotation should be on the same basis as those supplied by other firms. Its adjustment in the final account can then be treated in exactly the same way as described above.

6. Contractors should pay the nominated subcontractors the amount included in the engineer's certificate. Only agreed amounts for set-off, i.e. main contractor's charges against a subcontractor, can be legitimately deducted.

APPENDIX 7

DAYWORK ACCOUNTS

Daywork is construction work that is paid for on a time and materials basis. This basis of payment is described in a daywork schedule included within the contract documents. It is usual to adopt the 'Schedule of Dayworks carried out incidental to Contract Work' issued by the Federation of Civil Engineering Contractors (FCEC). Contractors are invited to include a percentage addition for overheads and profit on daywork accounts within the tender documents. The percentages applied on labour, materials and plant costs are different and are typically 120 per cent, 10 per cent and $12\frac{1}{2}$ per cent respectively.

Daywork is applied to work that cannot be properly measured or valued accurately within the terms of the contract. It frequently occurs because the work being executed is carried out under different conditions and circumstances. For example, a contractor has completed a paved area of work satisfactorily. Upon its completion instructions are received from the engineer to make changes by providing additional drainage. This involves excavating, constructing the drainage and making good. It would be unfair under these circumstances to reimburse the contractor at normal bill rates. Daywork rates are therefore used. Dayworks often arise from an engineer's instruction.

Principles

1. The contractor should inform the engineer that some work is intended to be carried out on a daywork basis.
2. Each daywork should be separately identified by a reference number and relate to a specific engineer's instruction.
3. Each daywork sheet should accurately describe the work that has been carried out.
4. Daywork sheets should be submitted on a regular basis to the engineer for signature. This signature is for record purposes only and indicates that the work has been properly executed and that information is correct, i.e. labour hours, material quantities, plant hours.
5. Daywork sheets are often used to record items of work that are being executed uneconomically. This partially reflects the nature of some site

instructions. It may, for example, include using materials in small quantities, abnormal handling charges, using plant inefficiently.

6. Times must be allowed on daywork sheets for clearing away rubbish and protecting the finished work, for access problems, etc. Sometimes, because of the nature of the work involved, these are higher than normal.

7. Subcontractors must prepare their daywork sheets in a similar manner.

8. When the sheet has been signed it should be filed along with the relevant instruction, drawings, details, sketches, delivery notes, etc. to be used for future reference purposes.

9. The submission of a daywork sheet does not mean that the work will be paid for on this basis. This is a separate matter for negotiation between the engineer and the contractor.

DELAYS

Delays to the regular progress of the works can be divided into four separate groups:

- by the contractor
- by the employer (or those employed directly by the engineer)
- by the engineer
- beyond the control of any party involved

The contractor

The contractor is responsible for completing the works within the time available for completion and in accordance with the programme approved by the engineer. The contractor must start the works as soon as possible and then proceed with due expedition and without delay to complete the works. The contractor must with due diligence proceed with the works or otherwise will be in breach of the obligations under the contract.

Where the contractor fails to complete the works (or part of the works where they are divided into sections) on time, the employer's remedy is to be awarded liquidated damages, under clause 47, and in accordance with the amounts stated in the Appendix to the conditions of contract.

The employer or engineer

Where a delay results from actions or inaction on the part of the employer or engineer, the contractor's remedy is to be awarded an extension of time under clause 44 of the conditions of contract. The effect of awarding the contractor an extension of time is twofold:

- removal of the option to apply liquidated damages
- possibility of a claim for loss and expense

Delays which are the responsibility of the employer or engineer include:

- adverse physical conditions (clause 12)
- late issue of drawings and specifications (clause 7)
- late issue of instructions (clause 13)

- failure to give timely possession of the site (clause 42)
- providing facilities for other contractors (clause 31)
- unforseeable design criteria (clause 14)
- unreasonable delays in engineer's consent (clause 14)

Many of the above are common occurrences with all types of construction work and provision therefore exists in all forms and conditions of contract.

The contractor may consider that the following are reasons to be granted an extension of time under clause 44 of the conditions of contract:

- variations ordered under clause 51 (1)
- increased quantities referred to in clause 51 (4)
- any cause of delay referred to in the conditions of contract
- exceptional adverse weather conditions
- other special circumstances

Beyond the control of any party involved

Because of the way that construction works are carried out, there are occasions where delays are caused without either party to the contract being at fault. Provisions exist in construction contracts to grant an extension of time to cover these situations. The ICE Conditions of Contract describe these as 'other special circumstances of any kind whatsoever which may occur'. In other forms of contract, for example, the Joint Contracts Tribunal Form of Contract, 1980, it is termed *force majeure*. The precise meaning of this term is not clear. It is frequently used with reference to all circumstances independent of the will of man and which are not in his power to control.

The inclusion of such clauses essentially limits the contractor's risk in contract by removing the liability to pay damages for late completion when such an event occurs. If such clauses were not included, then, through the fault of neither party, it could be impossible for one party to meet its obligations. The principles are therefore as follows:

- the party that is prevented from carrying out its work will be granted an extension of time
- the contract continues and remains in force and all other obligations remain

CLAIMS

The ideal civil engineering contract is where all of the information is completed prior to the contractor starting work on site. In practice, this is rarely the case and some changes or variations to the work become necessary due to design changes or site conditions.

Some will argue that many civil engineering contractors begin to consider the implications of claims at the same time as they are preparing their tenders. Claims will not, however, correct incorrect decisions made at the tender stage such as:

- incorrect tendering and estimating
- misunderstanding the complexities of the project
- poor perceptions of market conditions

Nor are claims the solution to poor site management of the construction process or to inadequate or inappropriate cost control procedures. It is not the purpose of the employer to reimburse the contractor's own deficiencies, nor can it be assumed that where an experienced contractor makes a loss on a project that this will necessarily result in a claim.

Legitimate claims are based upon contractual rights and must therefore be attributed to specific contractual clauses and events. Whilst a contractor may have such claim, it is also unlikely that the fault will be one-sided. The amount paid for a contractual claim is therefore frequently less than the amount submitted in a statement by the contractor. Claims are the subject of negotiation and agreement between the parties concerned.

Reasons for claims

Contractual claims occur for a variety of different reasons. The main reasons are as follows:

- variations to the original contract caused by changes to the design or specification or changes necessitated by unforeseen ground conditions
- late receipt of instructions from the engineer
- postponement of work
- poor co-ordination of the project by the engineer

- frequent changes causing disruption to the contractor's programme
- changes in the nature of the works

The initiative for contractual claims inevitably comes from the contractor. Where a claim is being considered, the contractor should inform the engineer of the circumstances involved. Adequate records can therefore be made and these can be verified by the engineer as true and correct. Where appropriate, the contractor should take steps to reduce the effects of a claim and the loss that might be incurred by the employer.

Common types of claim

The following represent some of the more common types of claim submitted by contractors:

- increased costs of preliminary items due to changes to the contract programme
- uneconomic use of mechanical plant or labour
- disruption of the regular progress of the works through the issue of site instructions
- site conditions different from those indicated in the contract documents
- standing or waiting time by mechanical plant or labour
- execution of work at a later stage than planned
- additional contract overheads or on-costs
- acceleration of the contract programme

Preparation of the claim

This is a combination of collecting the relevant facts surrounding a particular difficulty, writing a report which explains the rationale for the claim and calculating the costs involved, over and above those paid through remeasurement of the works.

Ex-gratia payments

These are payments made by an employer to a contractor for goodwill. They have no contractual basis or explanation and are not legally enforceable. They occur where a contractor has satisfactorily completed a project but has made a loss on the contract. The ex-gratia payment is made to off-set this loss. These are not common occurrences.

APPENDIX 10

FINAL ACCOUNTS

The ICE Conditions of Contract require the contractor to submit to the engineer a statement of final account and supporting documentation (clause 60.4). This will show in detail the value in accordance with the contract. It includes the total amount due to the contractor up to the date of the defects correction certificate. Within three months after receiving this information the engineer should issue a certificate stating the amount which is finally due under the contract from the employer to the contractor. The amount due is less previous payments that have been made by the employer to the contractor. The amount should be paid within 28 days of the engineer's certificate. The only adjustment from this amount by the employer is for any liquidated damages, under clause 47, due to late completion of the works.

Prior to preparing the final account it will be necessary to collect and cross-reference all the necessary documentation such as engineer's instructions, site meeting minutes, daywork sheets, nominated subcontractors' accounts, etc. Figure A10.1 is a simplified version of a summary for a final account.

In addition the engineer may have agreed interest on overdue payments as clause 60 (7) at a rate of 2 per cent above the bank rate.

Final account

Remeasurement of works.............................

Daywork account..

Nominated subcontractors..........................

Contractor's labour and other charges...........

Contract price fluctuations...........................

Contractual claims.......................................

                  Total.................

Less amount of interim payments......................

                  Total.................

Balance outstanding.....................................

Figure A10.1 Simplified version of a summary for a final account

APPENDIX 11

INSOLVENCY

Insolvency is a generic term that covers both individuals and companies. Bankruptcy applies specifically to the insolvency of individuals and liquidation to companies.

Before a contractor goes into liquidation, there are often signs that the firm is in financial difficulties. These may include:

- non-payment of subcontractors
- requests for earlier interim payments
- reduction in the progress of the works due to a reduced labour force or lack of material deliveries
- staffing changes
- rumours, which may be unfounded

When liquidation occurs, it often happens very quickly since the firm will have exhausted its last resources in an attempt to forestall this situation. It will be necessary to work strictly to the conditions of contract and to ensure that:

- plant and materials are not removed from the site
- insurances remain effective
- payment has been made to subcontractors
- defective work is being rectified

It will also be necessary to act quickly to:

- increase site security
- stop further payments to the contractor
- attempt to maintain essential services on site
- keep the bond holder informed
- ensure that insurances remain effective
- protect the works from damage

It may be necessary to make photographic records of the works at the time of the liquidation and to consider ways of completing the project. The effects of liquidation may not necessarily terminate the contract, since the appointed liquidator has the power, after obtaining leave, to carry on the business of the debtor for the beneficial winding up of the company. Projects that are well advanced may be profitable and are more likely to be completed in accordance with the contract than projects that are in the early months of their life. The completion of the project by the liquidator is likely to be the employer's favoured course of action.

Factors to consider

- Materials: Those which have been paid for in interim certificates are the employer's, unless their title is defective. Materials on site that have not been paid for will be used to complete the works. Payment will not be made at this stage, but their costs will be taken into account in the final account.
- Plant and temporary works: These will be used by the employer to complete the contract.
- Retention: This is one of the main buffers which the employer has to off-set against any additional charges.
- Subcontractors: There is no legal obligation to pay firms twice where payments have been made to the contractor but not passed on to subcontractors. However, in order for these firms to continue working this may be necessary and then to reduce the amounts owed to the contractor in the final account.

Completion of the project

It will be necessary to arrange a meeting between all those concerned, including the liquidator and the bond holder. Since the original contract was probably awarded to the lowest bidder, appointing a new firm may immediately increase the overall costs, ignoring the costs associated with its administration. It will be necessary to determine the costs that should have been incurred had the project been constructed as originally intended. The employer's loss may include:

- delay in completion
- additional payments to secure completion
- temporary site protection
- additional site security
- additional insurances
- extra professional fees
- double payments (to subcontractors)

Financial situation

Example:

Amount of original contract	10,000,000
Agreed variations, PC sums, dayworks increased costs, claims priced at the original contractor's rates	340,000
Amount of final account from original contractor	10,340,000

Amount paid to the original contractor after retention deductions	5,670,000	
Amount of completion contract	4,950,000	
Agreed variations, etc.	220,000	
Additional professional fees	<u>106,000</u>	<u>10,946,000</u>
Debt owed by liquidator to employer		<u>606,000</u>

If there was a performance bond for 5 per cent of the contract sum, then the employer should receive £500,000 towards this amount. If a final dividend, payable at some time later, from the liquidator was announced at 5p in the £1, then the employer would receive a further £30,300. The ultimate loss to the employer is then calculated as £75,700. No further payments will be made to the original contractor. Retention monies owed to the first contractor are used in the above calculation to offset the increased costs of employing a new contractor to complete the works.

INDEX